GOODS OF THE MIND, LLC

Math Challenges
for
Gifted Students

PRACTICE TESTS IN MATH KANGAROO STYLE FOR STUDENTS IN GRADES 1-2

Cleo Borac, M. Sc.
Silviu Borac, Ph. D.

This edition published in 2014 in the United States of America.

Editing and proofreading: David Borac, M. Mus.
Technical support: Andrei T. Borac, B.A., PBK

Send all inquiries to:

Goods of the Mind, LLC
1138 Grand Teton Dr.
Pacifica
CA, 94044

Math Challenges for Gifted Students
Level I (Grades 1 and 2)
Practice Tests in Math Kangaroo Style for Students in Grades 1-2

Contents

1 Foreword 5

2 Test Number One 6

3 Test Number Two 15

4 Test Number Three 23

5 Test Number Four 31

6 Test Number Five 41

7 Test Number Six 51

8 Hints and Solutions for Test One 61

9 Hints and Solutions for Test Two 65

10 Hints and Solutions for Test Three 71

11 Hints and Solutions for Test Four 76

12 Hints and Solutions for Test Five 83

13 Hints and Solutions for Test Six 90

Foreword

This workbook contains six exams that are similar to the Math Kangaroo contest, an international mathematics competition for students in grades 1-12.

As any contest paper, the difficulty of the items is staggered. The 3-point problems are relatively easy problems based on observation, elementary counting, and reading comprehension. The 4-point and the 5 point are problems that require more creative application of the concepts studied in school at the specific grade level.

The authors recommend this book as an additional study material to the series "Competitive Mathematics for Gifted Students" - level 1. As the student progresses through the material of the series, these tests are useful for assessment as well as for training specific competitive skills such as: time management, stamina, and focusing over a longer period of time. We recommend taking one of these tests every month or so. The student should have 75 minutes of contiguous time to solve without using a calculator. Using scratch paper is strongly suggested. The student should make diagrams, tables, and show work for each problem.

The authors are grateful to Maya Abiram for her help in reviewing the material. Maya is an avid problem solver who has won national level awards at Math Kangaroo each year since 2010 when she was in second grade.

Last but not least, these problems are good fun!

Test Number One

3-point problems

1. Counting down from 39 (39 is the first number), what is the 12th number?

   (A) 26 **(B)** 27 **(C)** 28 **(D)** 41 **(E)** 42

2. A bridge is 250 feet long. A man can cross it if he takes 250 steps. A horse can cross it if it makes 50 steps. How many steps are needed to cross the same bridge if the man rides the horse?

   (A) 5 **(B)** 50 **(C)** 100 **(D)** 200 **(E)** 250

3. In the figure, the fruit must be placed in a line so that, from left to right, their names are in alphabetical order. How many of them have not changed place?

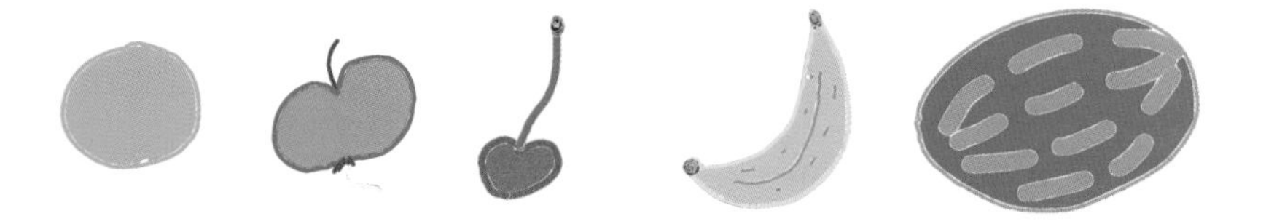

   (A) 0 **(B)** 1 **(C)** 2 **(D)** 3 **(E)** 4

4. Diana takes 35 minutes to get from home to school. If she left her home at 7:35 AM, what time did she arrive at school?

   (A) 7:65AM **(B)** 8:00AM **(C)** 8:05AM **(D)** 8:10AM **(E)** 8:15AM

5. What is the number that is larger by 7 than the number that is smaller than 125 by 11?

   (A) 114 **(B)** 118 **(C)** 121 **(D)** 129 **(E)** 132

6. The butterflies want to land on flowers with a matching number. Which flower will the green butterfly land on?

   (A) blue **(B)** yellow **(C)** orange **(D)** white **(E)** purple

7. Today is Saturday. What day of the week will it be when tomorrow will be "yesterday"?

   (A) Sunday **(B)** Monday **(C)** Tuesday **(D)** Thursday **(E)** Friday

8. Which of the choices is most likely to be a continuation of the pattern in the direction indicated by the arrow?

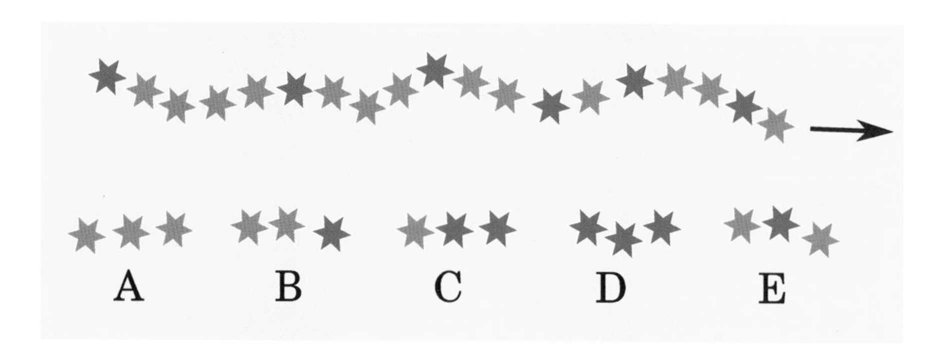

(A) A **(B)** B **(C)** C **(D)** D **(E)** E

4-point problems

9. By using a bar of soap for a week, the soap's weight decreases to half. How many months will it take to use up 12 bars of soap? Each month has 4 weeks.

(A) 4 **(B)** 5 **(C)** 6 **(D)** 8 **(E)** 24

10. Flora stands in a line for a movie ticket. She is fourth from the cashier and there are 8 people lined up behind her. How many people are standing in line at the moment?

(A) 10 **(B)** 11 **(C)** 12 **(D)** 13 **(E)** 14

11. Which operation must be done in block K to make the following true?

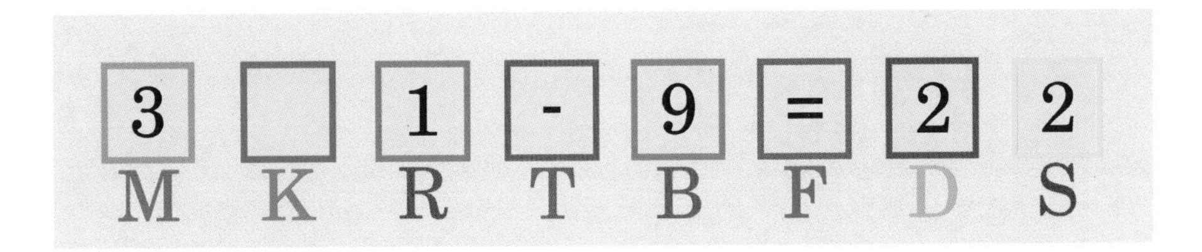

(A) +
(B) −
(C) ×
(D) =
(E) remove block K

12. Tom purchased 20 balloons for his birthday party and paid 2 dollars for them. While Tom and his father were decorating the room, some of the balloons popped. Tom had to pay 20 cents to replace them. How many balloons had popped?

(A) 1 **(B)** 2 **(C)** 3 **(D)** 4 **(E)** 5

13. Which of the loops will form a knot when you pull the ends?

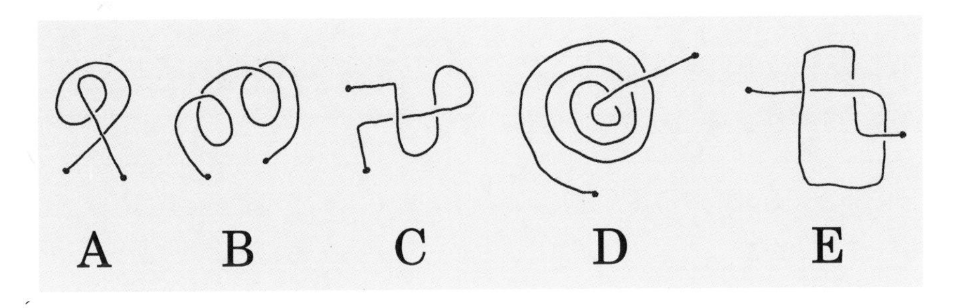

(A) A **(B)** B **(C)** C **(D)** D **(E)** E

14. How many of the following operations are impossible?

- cut a square into squares only
- cut a triangle into triangles only
- cut a circle into circles only
- cut a circle into triangles only

(A) 0 **(B)** 1 **(C)** 2 **(D)** 3 **(E)** 4

15. How many two digit numbers are there?

(A) 100 **(B)** 99 **(C)** 90 **(D)** 91 **(E)** 89

16. An office space has walls made of panels that divide it into rooms. At least how many of the 12 blue panels do we have to remove to make the following space have only 5 rooms?

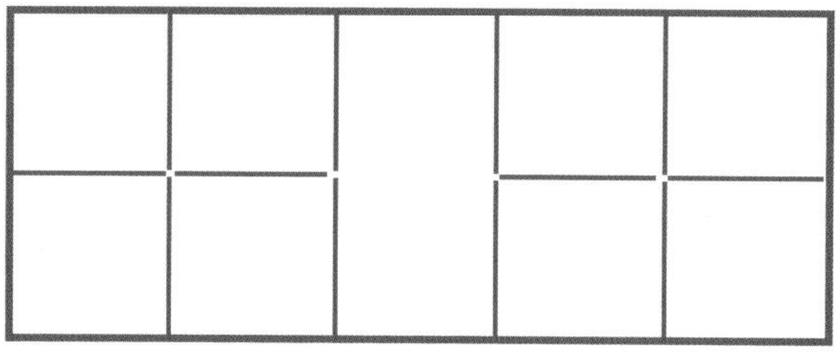

(A) 3 **(B)** 4 **(C)** 5 **(D)** 6 **(E)** 7

5-point problems

17. Daniella has the sum $1 + 2 + 3 + 4 + 5 = 15$. She wants to change one of the + (plus) operators with a − (minus) to obtain the result 14. How many different choices does she have?

(A) 0 **(B)** 1 **(C)** 2 **(D)** 3 **(E)** 4

18. Terence gives Donald 8 of his toys. Donald gives Terence 6 of his toys. Now, each of them has 13 toys. How many toys did Terence have to start with?

(A) 11 **(B)** 12 **(C)** 13 **(D)** 14 **(E)** 15

19. A square grid is formed of 9 small squares. Edwiga decides to paint some of the sides of the small squares in green. What is the smallest number of sides she must paint so that there no small square left that has the same color on all four sides?

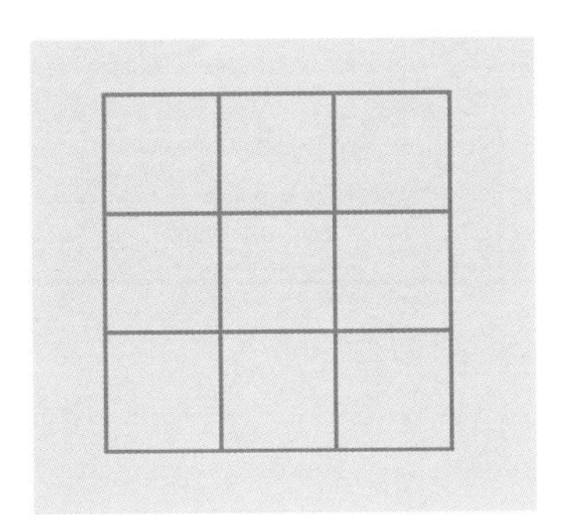

(A) 3 **(B)** 4 **(C)** 5 **(D)** 6 **(E)** 7

20. At a chess meet between Hill Middle School and Valley Middle School, the team from Hill won 4 games and the team from Valley won 5 games. There were no ties and each player played only one game. How many players from both schools participated in the tournament?

(A) 5 **(B)** 9 **(C)** 10 **(D)** 18 **(E)** 20

21. Puss in Boots can jump one unit at a time either up, down, left, or right on the square grid. He is very hungry and wants to get a mouse as quickly as possible. Which mouse will he try to catch?

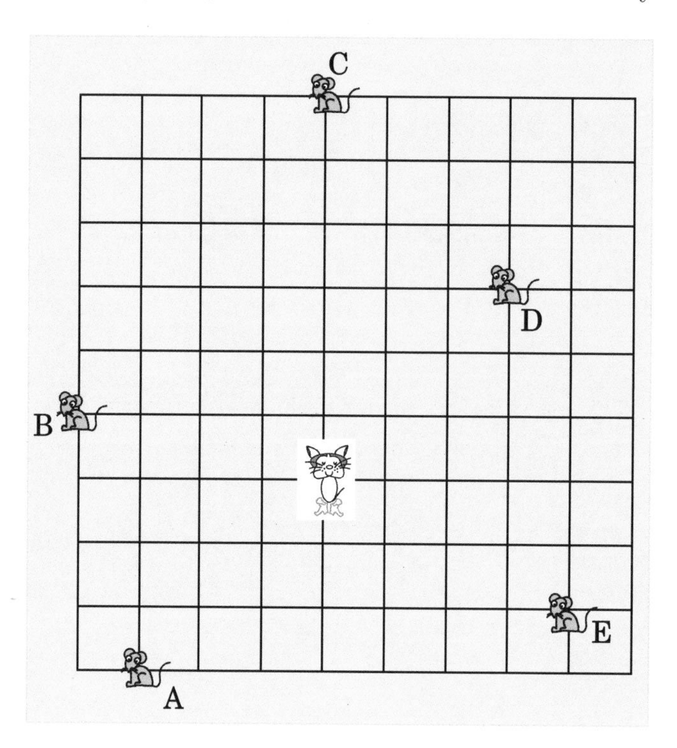

(A) *A* **(B)** *B* **(C)** *C* **(D)** *D* **(E)** *E*

22. Anda and Stella are practicing at the ballet studio. They start at opposite positions and walk towards one another. When their positions are exchanged, they turn around and do the same again. They keep doing this exercise for 20 minutes and ended up where they had started. Approximately, how many minutes have they been able to see each other?

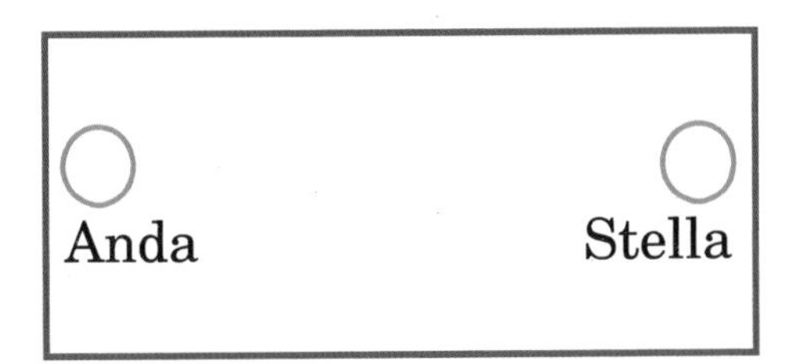

(A) 5 **(B)** 10 **(C)** 15 **(D)** 20 **(E)** 40

23. Erica lives on the middle floor of an 11-story building. Erica is going up in an elevator that has stopped on the second floor. How many more floors does the elevator have to climb to bring Erica on her floor?

(A) 3 **(B)** 4 **(C)** 5 **(D)** 6 **(E)** 7

24. Two horses, Lonely and Valentine, participated in a race. Lonely crossed the arrival line 4 minutes before Valentine. If both horses would have run at half the speeds they actually ran at, how many minutes before Valentine would Lonely have arrived?

(A) 1 **(B)** 2 **(C)** 4 **(D)** 6 **(E)** 8

Answer Key for Test One

3-point problems	4-point problems	5-point problems
1. C	9. C	17. A
2. B	10. C	18. E
3. C	11. E	19. C
4. D	12. B	20. D
5. C	13. E	21. B
6. E	14. C	22. B
7. B	15. C	23. B
8. B	16. B	24. E

Test Number Two

3-point problems

1. How many chopsticks are there for each saucer?

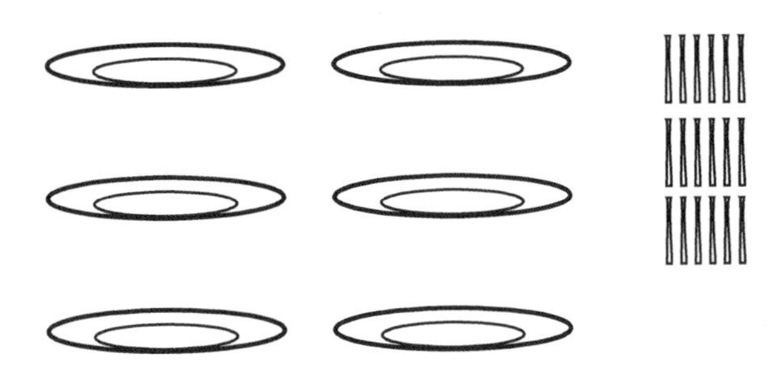

(A) 1 **(B)** 2 **(C)** 3 **(D)** 4 **(E)** 5

2. Which green piece can be combined with the orange piece to form a square?

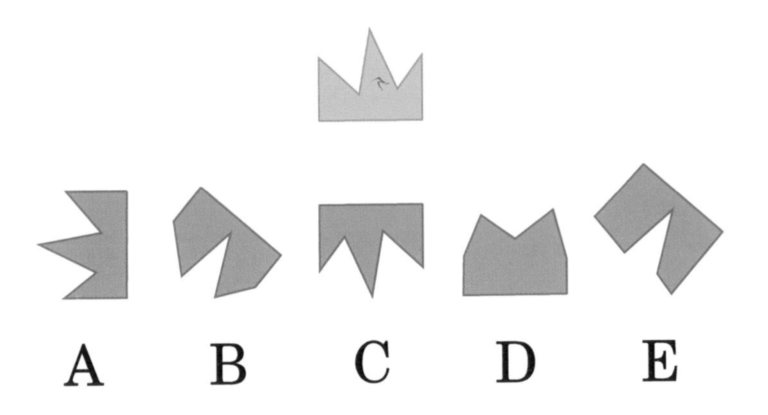

(A) A **(B)** B **(C)** C **(D)** D **(E)** E

3. Alena has a stack of many red and green cards. What is the largest number of different cards she can make if she writes an even digit on each card?

(A) 4 **(B)** 5 **(C)** 8 **(D)** 10 **(E)** 18

4. How many pairs of wings do the animals in the picture have?

(A) 4 **(B)** 5 **(C)** 8 **(D)** 9 **(E)** 10

5. Manuel makes a pattern with animal and plant cards. His pattern starts from the left and continues towards the right like this:

From left to right, the 11-th card will depict:

(A) a bird **(B)** an insect **(C)** a flower **(D)** a mammal **(E)** a reptile

6. Alan has pressed at the same time two of the keys on his phone that have numbers on them. If he adds the two numbers, what is the largest sum he could get?

   (A) 10 **(B)** 15 **(C)** 16 **(D)** 17 **(E)** 18

7. In Rainville, if you say "It has rained yesterday," it is true on any Thursday and Sunday. If you say "It will rain tomorrow," it is going to come true on any Monday or Thursday. On which of the following days it is possible to have a sunny day in Rainville?

   (A) Sunday **(B)** Saturday **(C)** Friday **(D)** Wednesday **(E)** Tuesday

8. Krista follows a trail of even numbers. She can only move left, right, up, or down. Which house does she reach?

   (A) *A* **(B)** *B* **(C)** *C* **(D)** *D* **(E)** *E*

4-point problems

9. Five boys and five girls stand in a circle. Each of them holds hands with their neighbors. How many pairs of hands are formed?

   (A) 5 **(B)** 10 **(C)** 15 **(D)** 20 **(E)** 25

10. Due to the time difference, it is 5:10 PM in Midville when it is 3:10 PM in Westville and when it is 6:10 PM in Eastville. What is the time in Eastville when it is 6:15 AM in Westville?

    (A) 3:15 AM **(B)** 4:15 AM **(C)** 5:15 AM **(D)** 7:15 AM **(E)** 9:15 AM

11. Each of my three sisters have two brothers. Each of my brothers has four sisters. How many siblings are we?

    (A) 5 **(B)** 6 **(C)** 7 **(D)** 8 **(E)** 9

12. Pippi has 4 pairs of red socks and 3 pairs of blue socks. At most how many pairs of socks with one blue and one red sock can she make?

    (A) 3 **(B)** 4 **(C)** 5 **(D)** 6 **(E)** 7

13. On a tree there are three families of birds. Each family has a nest with two chicks in it. In the morning, each father bird has provided a fresh worm for each chick and each mother bird in his family. How many worms did the three fathers bring today?

    (A) 6 **(B)** 8 **(C)** 9 **(D)** 10 **(E)** 12

14. Jared has 12 crickets more than his friend Fongo. If Jared gives 12 crickets to Fongo then:

    (A) Jared and Fongo have the same number of crickets.
    (B) Fongo has 12 more crickets than Jared.
    (C) Fongo has 6 crickets more than Jared.
    (D) Jared has 6 crickets more than Fongo.
    (E) It is not possible to determine who has more crickets.

15. The Golden Goose lays a golden egg each day. Simpleton collects the eggs each week and sells them at the market on Fridays. This Friday, he was offered either two chicks or three ducklings for each egg. If Simpleton decided to exchange all the eggs and got only 8 chicks, how many ducklings did he get?

    (A) 3 **(B)** 6 **(C)** 7 **(D)** 9 **(E)** 12

16. The day after tomorrow is how many days after the day before yesterday?
    (A) 2 **(B)** 3 **(C)** 4 **(D)** 5 **(E)** 6

5-point problems

17. Steven added two whole numbers and got 2015. Then, how many of the following statements must always be true?

    - one of the numbers is even
    - one of the numbers is smaller than the other
    - one of the numbers is smaller than 1000
    - both numbers are larger than 1000

    (A) 0 **(B)** 1 **(C)** 2 **(D)** 3 **(E)** 4

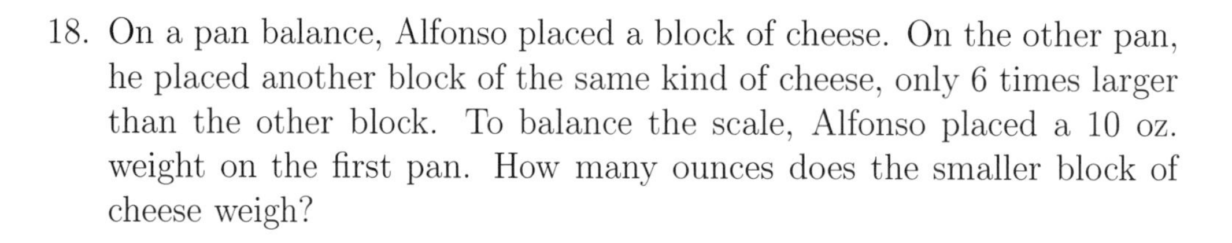

18. On a pan balance, Alfonso placed a block of cheese. On the other pan, he placed another block of the same kind of cheese, only 6 times larger than the other block. To balance the scale, Alfonso placed a 10 oz. weight on the first pan. How many ounces does the smaller block of cheese weigh?

    (A) 1 **(B)** 2 **(C)** 3 **(D)** 4 **(E)** 5

19. Jiminy says the alphabet backwards. The first letter is Z and the second is Y. At which count does he say the first vowel?

    (A) 5 **(B)** 6 **(C)** 7 **(D)** 11 **(E)** 12

20. Greta, Alma, Selma, Thera, and Jillian have been assigned a book to read. Greta finished reading her copy after Alma finished hers but before Selma finished hers. Thera finished after Alma but before Jillian. Jillian finished before Greta. Who was the slowest reader?

    (A) Alma **(B)** Greta **(C)** Jillian **(D)** Selma **(E)** Thera

21. Akhil has two boxes and three keys. He knows that the keys to the boxes are among the three. In the worst case, how many times does he have to try a key to a lock before he can tell which is the key that does not fit either lock?

    (A) 1 **(B)** 2 **(C)** 3 **(D)** 5 **(E)** 6

22. Daniel has 3 cards printed with **A** and 2 cards printed with **B**. He wants to use these cards to make words that read the same forwards as backwards. An example of such a word is **XZYZX**. How many different such words can he make using all his cards for each word?

    (A) 0 **(B)** 1 **(C)** 2 **(D)** 3 **(E)** 6

23. Zenovia added two numbers. Gina subtracted the same numbers and got the same result as Zenovia. One of the numbers must be:

    (A) even **(B)** odd **(C)** neither even nor odd
    (D) very large **(E)** equal to the other number

24. Five girls stand in a line. Alma, the tallest, stands on the left. She is followed by the other girls in decreasing order of their height: Greta, Jillian, Selma, and Thera. Thera is the shortest and is placed at the rightmost end. The girls plan to swap positions in pairs until they reverse the order: with Thera on the left and Alma on the right. At least how many swaps are needed?

    (A) 1 **(B)** 2 **(C)** 3 **(D)** 4 **(E)** 5

Answer Key for Test Two

3-point problems	4-point problems	5-point problems
1. C	9. B	17. C
2. B	10. E	18. B
3. D	11. B	19. B
4. A	12. D	20. D
5. A	13. C	21. C
6. D	14. B	22. C
7. A	15. D	23. A
8. E	16. C	24. B

Test Number Three

3-point problems

1. In the word ABRACADABRA, how many times is the letter A used?

(A) 2 **(B)** 3 **(C)** 4 **(D)** 5 **(E)** 6

2. Which basket has one less egg than the basket with a bow?

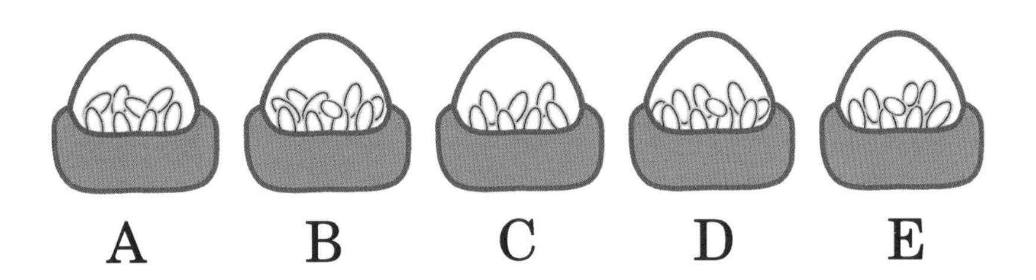

(A) A **(B)** B **(C)** C **(D)** D **(E)** E

3. Fox follows a path among trees and marks with a dot of paint all the trees that are on his path. How many trees are left unmarked?

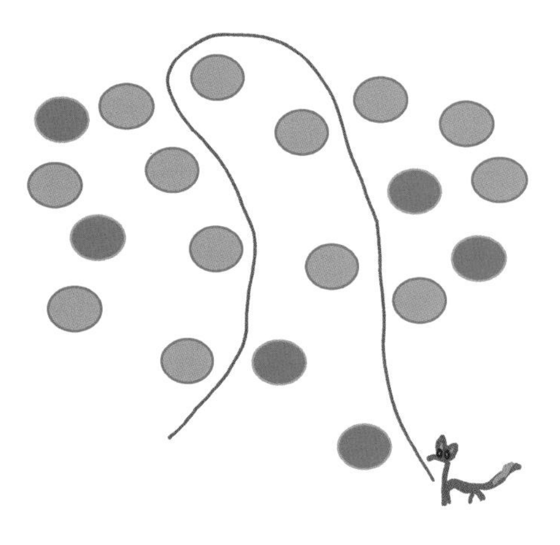

(A) 5 **(B)** 6 **(C)** 7 **(D)** 8 **(E)** 9

4. Alexandra counts aloud by 7s: 7, 14, 21, ... What is the fifth number she says?

(A) 28 **(B)** 34 **(C)** 35 **(D)** 36 **(E)** 37

5. On Mathymail street, the mail carrier puts 3 letters in each mailbox for the houses with even numbers and 2 letters in each mailbox for the houses with odd numbers. How many letters did he distribute?

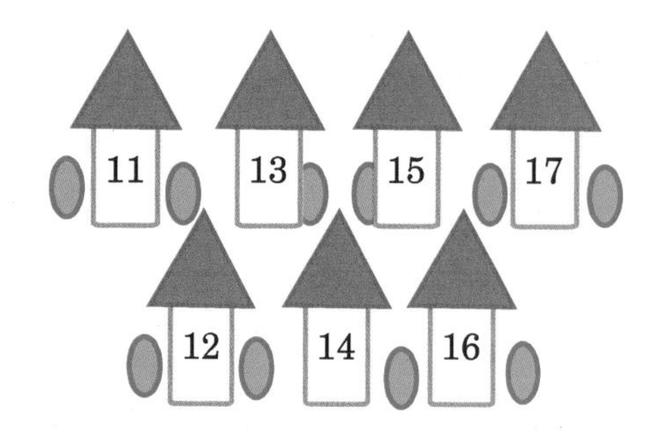

(A) 14 **(B)** 15 **(C)** 16 **(D)** 17 **(E)** 18

6. A car travels 40 miles in one hour without changing its speed. How many miles did it travel in 15 minutes?

   (A) 5 **(B)** 8 **(C)** 10 **(D)** 15 **(E)** 20

7. If every cat has two friends who are mice, what is the smallest number of mice that can be friends with 3 cats?

   (A) 2 **(B)** 3 **(C)** 4 **(D)** 5 **(E)** 6

8. Which of the choices represents a mirror that is placed behind the cards?

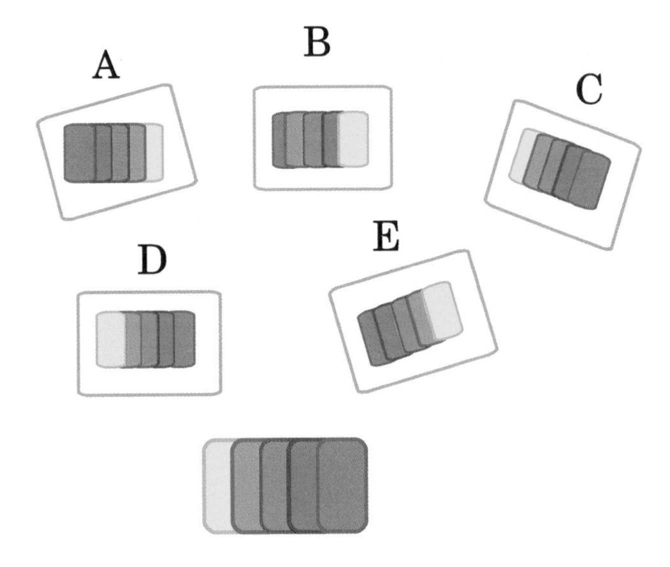

   (A) A **(B)** B **(C)** C **(D)** D **(E)** E

4-point problems

9. A rectangle is made out of three squares. Harry wants to make a larger rectangle by surrounding this rectangle once with squares of the same size. How many squares does he need?

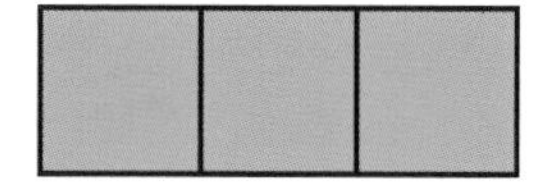

(A) 6 **(B)** 8 **(C)** 9 **(D)** 10 **(E)** 12

10. It is now 3:00 PM. If one hour would pass, then we would be 35 minutes late for our dance class. What time does the class start?

(A) 3:35 PM **(B)** 2:35 AM **(C)** 3:25 PM **(D)** 4:25 PM **(E)** 2:25 PM

11. Stan has 5 red triangles, 8 yellow triangles, and 9 green triangles. First, he paints 2 red triangles green, then he paints 4 yellow triangles red, and lastly he paints all the green triangles yellow. How many yellow triangles does he have now?

(A) 4 **(B)** 8 **(C)** 9 **(D)** 11 **(E)** 15

12. A digit can be written using some of the eight segments in the figure:

How many different digits are written using 5 of the segments?
(A) 0 **(B)** 1 **(C)** 2 **(D)** 3 **(E)** 4 5

13. Yolanda comes back from school at 4:00 PM on Mondays, Tuesdays, and Fridays. On Wednesdays and Thursdays, she comes back at 2:30 PM. Today she came back at 4:00 PM but tomorrow she will come back at 2:30 PM. What day of the week was the day before yesterday?

(A) Thursday **(B)** Friday **(C)** Saturday **(D)** Sunday **(E)** Monday

14. In how many ways can pairs of (non-negative) even numbers add up to 12?

(A) 1 **(B)** 2 **(C)** 3 **(D)** 4 **(E)** 5

15. A plane is flying 9000 feet above the ground. The clouds that can be seen through the windows are 7500 feet above the ground. How many feet above the clouds is the plane flying?

(A) 1000 **(B)** 1500 **(C)** 2000 **(D)** 2500 **(E)** 16500

16. Andrea and I were given a number. I added 5 to it and then subtracted 15. Andrea added 3 to it and then subtracted 13. The results we obtained differ by:

(A) 0 **(B)** 1 **(C)** 2 **(D)** 3 **(E)** 4

5-point problems

17. George battled the 3-headed dragon with his sword. If George cuts a head two more grow instead. After George cuts 11 heads, how many heads does the dragon have in total?

(A) 8 **(B)** 11 **(C)** 14 **(D)** 22 **(E)** 33

18. In the subtraction, which digit has been replaced by a triangle?

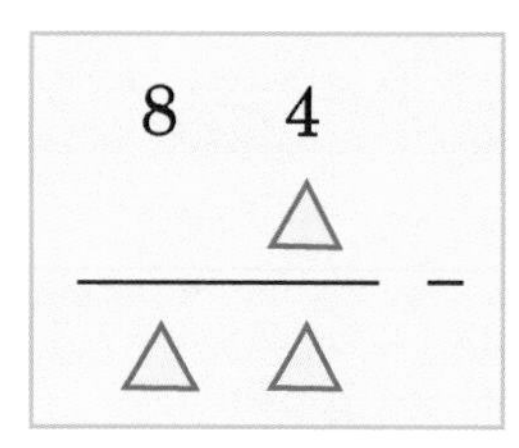

(A) 2 **(B)** 5 **(C)** 7 **(D)** 8 **(E)** 9

19. Andy ate 3 of Brad's biscuits. Brad ate 2 of Dan's biscuits. After Dan ate one of Andy's biscuits, each of them had 4 biscuits left. How many biscuits did Brad start with?

(A) 4 **(B)** 5 **(C)** 6 **(D)** 7 **(E)** 8

20. Five monkeys have 5 coconuts each. What is the smallest number of coconuts that some monkeys have to give other monkeys for the numbers of coconuts they each have to be consecutive numbers?

(A) 1 **(B)** 2 **(C)** 3 **(D)** 4 **(E)** 5

21. In a 4-digit number the first two digits add up to an even number and the last two digits add up to an odd number. The difference between the even and the odd number is 17. What is the first (leftmost) digit of the 4-digit number?

(A) 6 **(B)** 7 **(C)** 8 **(D)** 9 **(E)** there are multiple solutions

22. 3 glooks weigh as much as 2 maboons. 1 maboon weighs as much as 3 pogs. If we place one glook on a pan of a balance scale, how many pogs should we place on the other pan to balance the scale?

(A) 1 **(B)** 2 **(C)** 3 **(D)** 4 **(E)** 5

23. In how many ways can Anna, Chris, and Donald stand in a line, if they don't want to be ordered by height (neither increasing, nor decreasing)?

(A) 0 **(B)** 1 **(C)** 2 **(D)** 3 **(E)** 4

24. Dana is alone and makes games up for herself. She imagines that her toys live on a cube. She places each toy on the cube and sets the rule that each toy can only go forward. The sides of the cube are labeled: Front, Back, Left, Right, Up, and Down. The cube rests on the face called Down. Each time a toy visits a face for the first time, it gets a stamp in its passport. At most, how many different number of stamps can Dana's toys have in their passports?

(A) 1 **(B)** 2 **(C)** 3 **(D)** 4 **(E)** 5

Answer Key for Test Three

3-point problems		4-point problems		5-point problems	
1.	D	9.	E	17.	C
2.	C	10.	C	18.	C
3.	C	11.	E	19.	D
4.	C	12.	D	20.	C
5.	D	13.	D	21.	D
6.	C	14.	D	22.	B
7.	A	15.	B	23.	E
8.	D	16.	A	24.	D

TEST NUMBER FOUR

3-point problems

1. If Ildiko performs the following operations which result will she obtain?

$$6 + 6 - 11 + 6 + 6 - 11 =$$

(A) 1 **(B)** 2 **(C)** 3 **(D)** 4 **(E)** 5

2. How many balloons have the same number of dots on them?

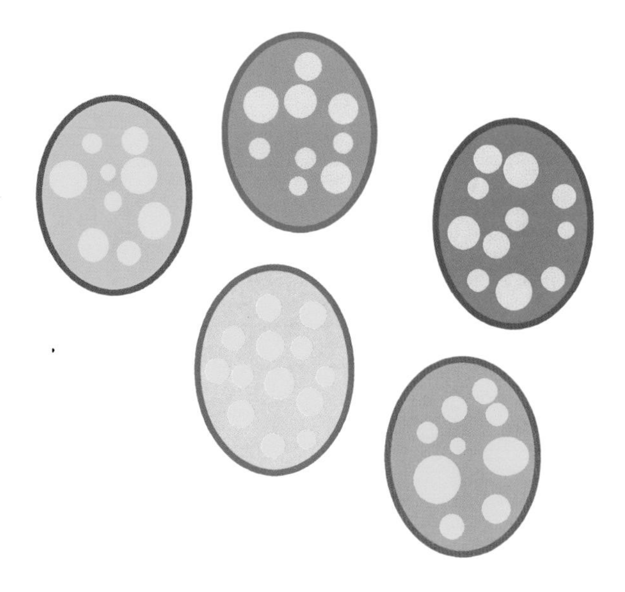

(A) 0 **(B)** 2 **(C)** 3 **(D)** 4 **(E)** 5

3. In the garden shed there are mom's gardening boots, dad's boots, Elmira's boots, and John's boots. How many boots are there in the shed in total?

   (A) 4 **(B)** 5 **(C)** 6 **(D)** 7 **(E)** 8

4. What is the next figure (to the right) in the pattern?

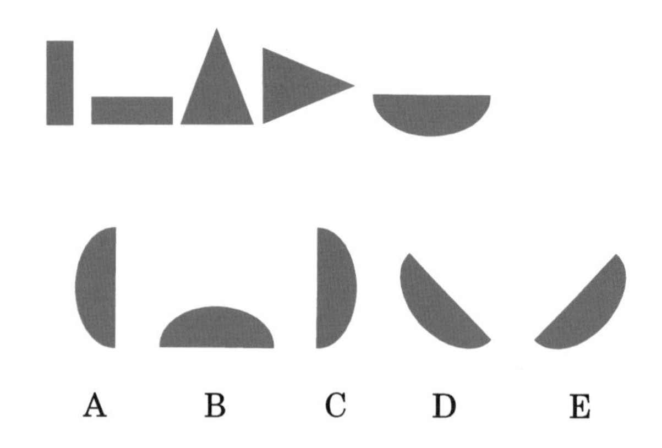

   (A) A **(B)** B **(C)** C **(D)** D **(E)** E

5. In a month there are 5 Thursdays. In the same month, there *cannot* also be:

   (A) five Mondays
   (B) five Tuesdays
   (C) five Wednesdays
   (D) five Fridays
   (E) five Saturdays

6. What is the least number of strokes one can use to cut this cake into seven slices?

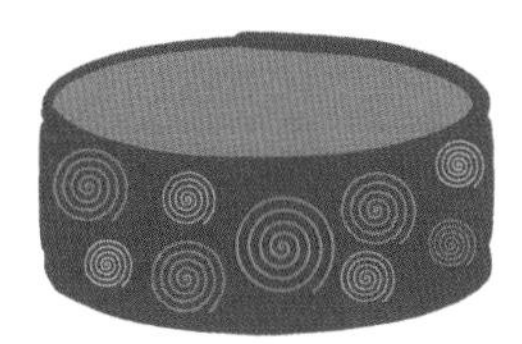

(A) 2 **(B)** 3 **(C)** 4 **(D)** 6 **(E)** 7

7. Andra uses a square stamp and a triangular stamp to make this pattern:

If she always holds the stamp in her right hand, at least how many times did she have to change the stamp she was using?

(A) 1 **(B)** 2 **(C)** 3 **(D)** 4 **(E)** 5

8. Jan has sticks that are 4 inches long, 6 inches long, and 3 inches long. How many different sizes of squares can he make if each square must be made out of exactly 8 sticks?

(A) 3 **(B)** 4 **(C)** 5 **(D)** 6 **(E)** 8

4-point problems

9. It is very cold today. The mother folds the blanket in three and wraps the baby twice around with it. How many layers of blanket cover the baby?

(A) 3 **(B)** 4 **(C)** 5 **(D)** 6 **(E)** 12

10. A train is 200 feet long. At 5:30 AM, the train started from the depot, where it had been parked for the night, and stopped at the first station to pick up passengers. If the engine traveled 3500 feet, how many feet did the last car travel?

(A) 3300 **(B)** 3500 **(C)** 3700 **(D)** 3900 **(E)** 7000

11. Some numbers are talking to one another:

    (a) - I am one less than four pairs.

    (b) - I am the largest digit.

    (c) - I am an even digit larger than 6.

    (d) - I am three more than two pairs.

    (e) - I am the second largest odd digit.

    How many different numbers have spoken?

    (A) 1 **(B)** 2 **(C)** 3 **(D)** 4 **(E)** 5

12. When we parked the car, I noticed it was the tenth car from the right as well as the thirteenth car from the left in its row. How many cars in total were there in the row?

(A) 13 **(B)** 20 **(C)** 21 **(D)** 22 **(E)** 23

13. A farmer sells cherries at the farmer's market. He has a balance scale with 2 pans and two weights: one of 2 lb and one of 4 lb. At least how many measurements does he have to make to weigh 3 lbs of cherries?

(A) 1 **(B)** 2 **(C)** 3 **(D)** 4 **(E)** 5

14. Three mothers and their five daughters attended a movie screening. At least how many persons were in this group?

    (A) 4 **(B)** 5 **(C)** 6 **(D)** 7 **(E)** 8

15. After the small circle does half a turn clockwise and the large circle does half a turn counter-clockwise, the arrow marked 4 points to the:

    (A) blue star
    (B) red star
    (C) moon
    (D) sun
    (E) constellation purple star

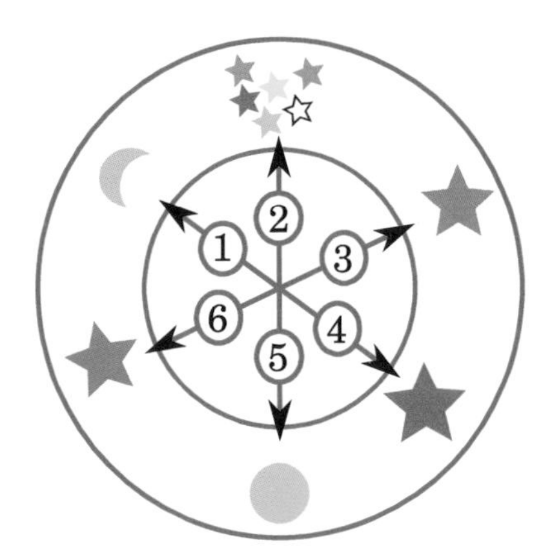

16. Randy has a card that is red on one side and blue on the other. The card faces with the blue side up. Randy turns it from side to side. How many times has Randy turned the card by the time the red face comes up the third time?

    (A) 3 **(B)** 4 **(C)** 5 **(D)** 6 **(E)** 7

5-point problems

17. Based on the figure with balloons, how many of the following statements are true?

    1. If a balloon is green, then it is small.
    2. If a balloon is large, then it is orange.
    3. If a balloon is small, then it is green.
    4. If a balloon is orange, then it is large.

(A) 0 **(B)** 1 **(C)** 2 **(D)** 3 **(E)** 4

18. Daniel has three more toys than Mark. Mark has five toys less than Shawn. If Shawn gives half of his toys to Mark, then Mark will have as many toys as Daniel. How many toys does Daniel have?

(A) 1 **(B)** 2 **(C)** 3 **(D)** 4 **(E)** 5

19. Andra makes a number with the digits 7, 7, and 8. Michael makes a number with the digits 7, 8, and 9. If they subtract the smaller number from the larger one, what is the smallest possible difference?

(A) 2 **(B)** 11 **(C)** 20 **(D)** 110 **(E)** 200

20. Dan has 5 cubes with faces colored in different colors. He is able to make a tower that is 5 cubes high by glueing together only blue faces. At least how many of the cubes must have at least 2 blue faces?

    (A) 1 **(B)** 2 **(C)** 3 **(D)** 4 **(E)** 5

21. The three bears have three baskets with apples in them. Mama Bear and Little Bear have as many apples as Papa Bear. Mama Bear has twice as many apples as Little Bear. Papa Bear gives Little Bear 3 apples and now they each have the same number of apples. How many apples do they have in total?

    (A) 9 **(B)** 12 **(C)** 15 **(D)** 18 **(E)** 30

22. Laura has two identical cards with the digit 6 on them:

How many different numbers can she make with the two cards?

(A) 1 **(B)** 2 **(C)** 3 **(D)** 4 **(E)** 6

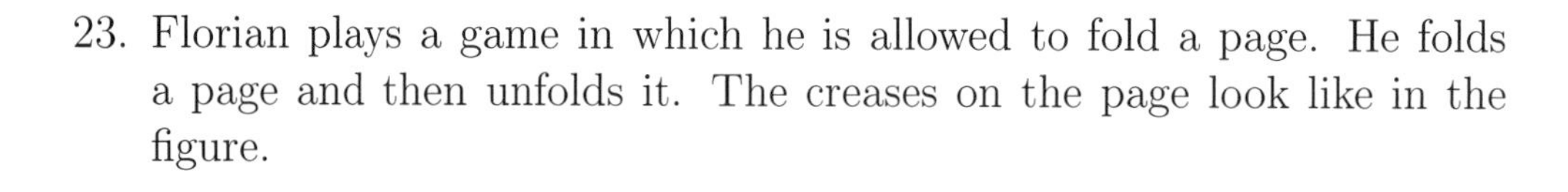

23. Florian plays a game in which he is allowed to fold a page. He folds a page and then unfolds it. The creases on the page look like in the figure.

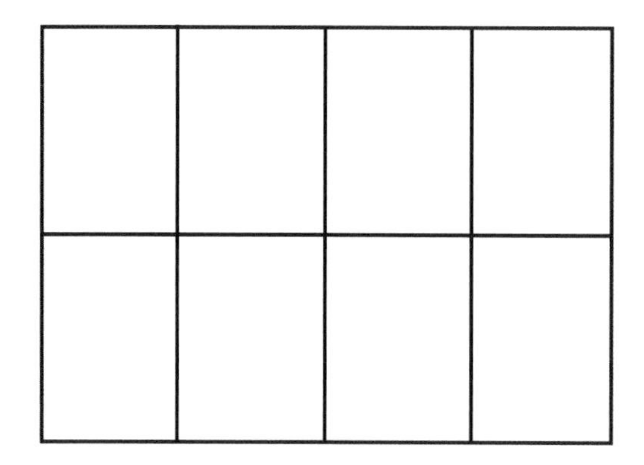

When he had finished folding it, how many layers of paper were there?

(A) 2 **(B)** 4 **(C)** 6 **(D)** 8 **(E)** 10

24. The Bear, the Wolf, the Fox, and the Rabbit are playing a game of cards. The game starts with the players exchanging cards with one another. Each pair of animals is allowed only one exchange. If Bear exchanged a total of 3 cards, Wolf exchanged 2 cards, and Rabbit exchanged only 1 card, how many cards did the Fox exchange?

(A) 1 **(B)** 2 **(C)** 3 **(D)** 4 **(E)** 5

Answer Key for Test Four

3-point problems	4-point problems	5-point problems
1. B	9. D	17. C
2. C	10. B	18. D
3. E	11. C	19. A
4. A	12. D	20. C
5. A	13. C	21. D
6. B	14. C	22. E
7. A	15. B	23. D
8. D	16. C	24. B

Test Number Five

3-point problems

1. How many of the flowers have the same number of petals as leaves?

(A) 1 **(B)** 2 **(C)** 3 **(D)** 4 **(E)** 5

2. In the following list of words, how many have exactly 3 vowels in their spelling?

TREE, LEAF, FLAMINGO, PEAR, QUINCE, HORSE, FAREWELL,

FAREWELL, DOUBLE, MICE, ROSIN

(A) 2 **(B)** 3 **(C)** 4 **(D)** 5 **(E)** 6

3. In spring, each day the sun rises 5 minutes earlier than the previous day. Today, the sunrise was at 6:30 AM. How many days will it take for the sunrise to happen one hour earlier?

(A) 10 **(B)** 12 **(C)** 15 **(D)** 20 **(E)** 30

4. John is getting ready to go hiking. He has to change into hiking clothes. He has to change his socks, pants, T-shirt, hat, and sweater. How many pieces of clothing did he have to change?

   (A) 3 **(B)** 4 **(C)** 5 **(D)** 6 **(E)** 7

5. A stone has broken the center of a stained glass window. How many triangles are missing from it?

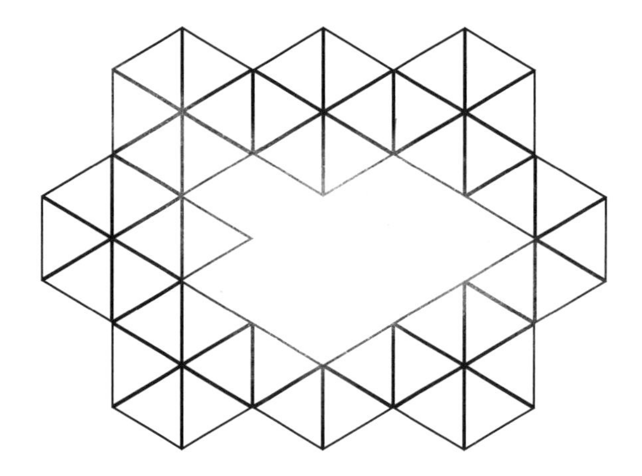

   (A) 12 **(B)** 13 **(C)** 14 **(D)** 15 **(E)** 16

6. In the Fairytale Garden there are 3 worms. Each worm is chased by a bird. Each bird is chased by 2 cats and each cat is chased by a dog. How many animals are there in this story?

   (A) 8 **(B)** 9 **(C)** 10 **(D)** 13 **(E)** 18

7. A clock chimes as many times as the hour, on each hour. It takes 1 second to chime a "ding" and 2 seconds to chime a "dong." If the clock chimes in a pattern of two dings followed by a dong, how many seconds does it take to chime 7 o'clock?

   (A) 7 **(B)** 8 **(C)** 9 **(D)** 10 **(E)** 12

8. Five people are leaving a restaurant. All of them had left their umbrellas at the check-in. Outside, it is raining, and they are rushing to catch the train. Alice grabs Ben's umbrella, Ben grabs Chloe's umbrella, Chloe grabs Daniel's umbrella, and Daniel grabs Alice's umbrella. How many of them left carrying their own umbrella?

   (A) 0 **(B)** 1 **(C)** 2 **(D)** 3 **(E)** 4

4-point problems

9. Andra and Tom are skipping stones on the surface of a lake. In their game, they get 1 point for each skip. If Andra had 3 stones that skipped 4 and 2 stones that skipped 3 and Tom had 4 stones that skipped 2 and 1 stone that skipped 6, by how many points was Andra's score larger?

   (A) 3 **(B)** 4 **(C)** 5 **(D)** 6 **(E)** 8

10. Five square pieces of cardboard are placed on a table. If they look as in the figure, in which order are they placed, from bottom to top?

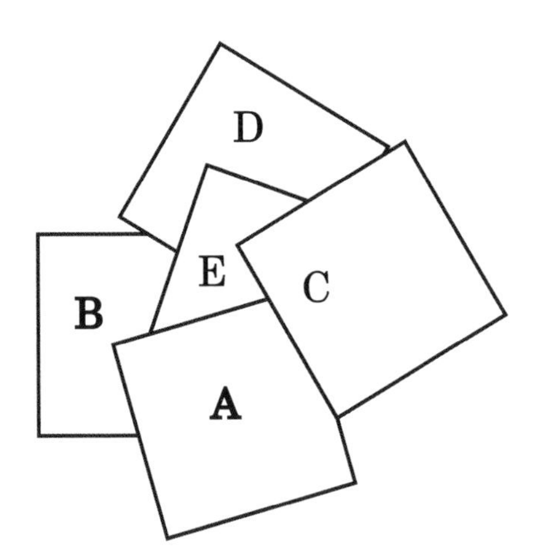

   (A) CAEDB **(B)** BEDAC **(C)** BAECD **(D)** BDEAC **(E)** BAEDC

11. Kro Frog and Kra Frog face each other and are 4 leaps apart. Kro takes 4 leaps in Kra's direction and Kra takes 4 leaps in Kro's direction. How many leaps apart are Kro and Kra now?

    (A) 0 **(B)** 2 **(C)** 4 **(D)** 8 **(E)** 12

12. Santa travels to his workshop from time to time to bring toys to his warehouse. He travels according to a pattern that starts like this:

    Thursday, Tuesday, Sunday, Friday, Wednesday

    What is the day of the week when Santa will travel next?

    (A) Sunday **(B)** Monday **(C)** Tuesday **(D)** Saturday **(E)** Wednesday

13. Bernard and Bianca are putting their sports equipment in the lockers. There is a row of 14 lockers. Bernard's locker is eighth from the left and Bianca's is eighth from the right. How many lockers are there between Bernard's locker and Bianca's locker?

    (A) 0 **(B)** 1 **(C)** 2 **(D)** 4 **(E)** 6

14. Norwegian Chef makes a fruit salad. For each two pineapples, he uses three apples and one pound of grapes. For tonight's party, he used three pounds of grapes. Which of the following computations will give us the total number of apples and pineapples he has used?

    (A) $2 \times 3 + 3 \times 3$ **(B)** $2 \times 3 \times 3 \times 3$ **(C)** $2 + 3 \times 3 + 3$ **(D)** $2 + 3 + 2 + 3$ **(E)** 5×2

15. Hedwig cut a square into rectangles like this one:

What is the smallest possible number of rectangles she has?

(A) 2 **(B)** 4 **(C)** 6 **(D)** 9 **(E)** 18

16. In the last month of March there were 12 days with temperatures of 48° F or less, and 16 days with temperatures 50° F or larger. How many days had temperatures between 48° F and 50°?

(A) 2 **(B)** 3 **(C)** 10 **(D)** 12 **(E)** 13

5-point problems

17. Martin steps on the scale holding a suitcase in his hand. The scale shows 160 lbs. Then, Martin picks up another suitcase and steps on the scale holding one suitcase in each hand. Now, the scale shows 180 lbs. If the two suitcases have the same weight, what is Martin's own weight?

(A) 40 **(B)** 60 **(C)** 120 **(D)** 140 **(E)** 160

18. Leo, Sandeep, Alvaro, Chloe, and Martin are building sandcastles at the beach. Leo builds a 10 inch high tower with a 2 inch deep cellar. Sandeep builds an 8 inch high tower with a 5 inch deep cellar. Alvaro builds a 12 inch high tower with a 1 inch deep cellar. Chloe builds an 11 inch high tower with a 2 inch deep cellar. Martin builds an 9 inch high tower with a 5 inch deep cellar. Whose building has the largest distance from the top of the tower to the bottom of the cellar?

(A) Leo **(B)** Sandeep **(C)** Alvaro **(D)** Martin **(E)** Chloe

19. Write the following numbers in increasing order:

$$55, 505, 5055, 5050, 5505, 50550, 55055, 50055, 50505$$

If you now want to place 50500 in the ordered list, it would be:

(A) fifth **(B)** sixth **(C)** seventh **(D)** eighth **(E)** ninth

20. Dana gave 5 pretzels to Jack, Jack gave 3 pretzels to Telly, and in the end they each had 6 pretzels. How many pretzels did Jack have to start with?

(A) 3 **(B)** 4 **(C)** 5 **(D)** 6 **(E)** 8

21. At a party, several pairs dance on the floor for 1 minute. After this, the pairs that danced take turns to execute a dance of their own for another minute each. The whole performance lasted 8 minutes. How many people have danced?

(A) 12 **(B)** 14 **(C)** 16 **(D)** 18 **(E)** 20

22. In the following addition, some of the digits have been replaced by geometric figures. Same figures cover same digits. What is the difference between the digits covered by the two different triangles?

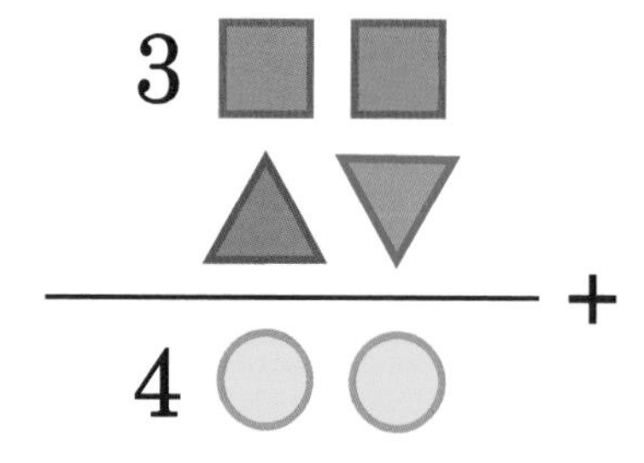

(A) 0 **(B)** 1 **(C)** 2 **(D)** 3 **(E)** 4

23. What is the smallest number of rooms Theseus must cross before he reaches the room of the Minotaur?

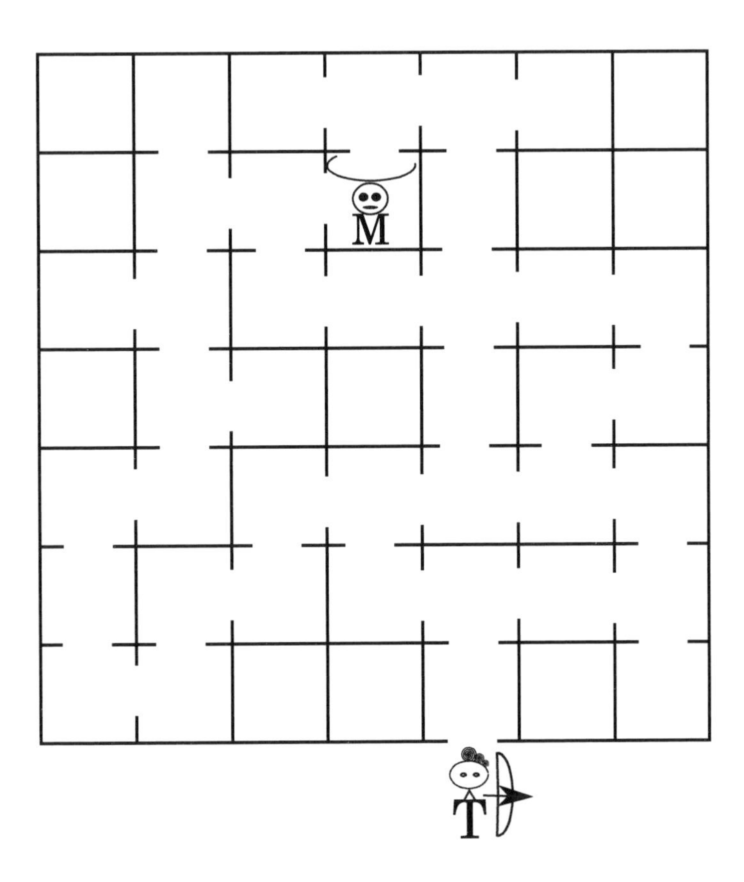

(A) 10 **(B)** 11 **(C)** 12 **(D)** 13 **(E)** 14

24. Andrew paints some of the squares in the 4×4 grid with blue, so that each white square has *exactly* 3 blue squares as neighbors. At most, how many white squares are there when he finishes painting?

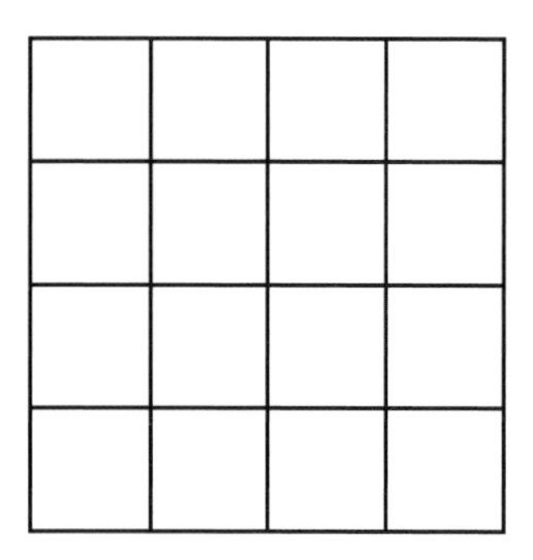

(A) 1 **(B)** 2 **(C)** 3 **(D)** 4 **(E)** 5

Answer Key for Test Five

3-point problems		4-point problems		5-point problems	
1.	C	9.	B	17.	D
2.	C	10.	B	18.	D
3.	B	11.	C	19.	C
4.	D	12.	B	20.	B
5.	C	13.	A	21.	B
6.	E	14.	A	22.	B
7.	C	15.	D	23.	C
8.	B	16.	B	24.	E

Test Number Six

3-point problems

1. Which of the answer choices represents a number larger than all other answer choices?

   (A) 898989 **(B)** 899899 **(C)** 898898 **(D)** 899889 **(E)** 898999

2. A balloon is hovering over a lake, 200 yards above the lake. The lake is 50 yards deep. A bag of sand is dropped from the balloon. How many yards long was the fall of the bag?

   (A) 150 **(B)** 200 **(C)** 250 **(D)** 1000 **(E)** it is impossible to determine

3. The analog clock face in the figure shows the current time. What color will the hour hand be on 8 hours from now?

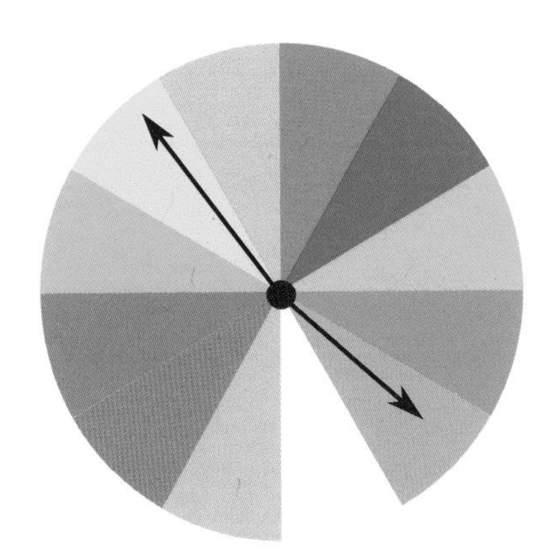

   (A) red **(B)** blue **(C)** yellow **(D)** green **(E)** grey

4. If a ticket to the zoo costs 5 dollars for a child and 10 dollars for an adult, how much do two parents with three children have to pay to buy their tickets?

   (A) 15 **(B)** 25 **(C)** 35 **(D)** 45 **(E)** 50

5. Five snails want to get to a leaf. Which one has the shortest path to go?

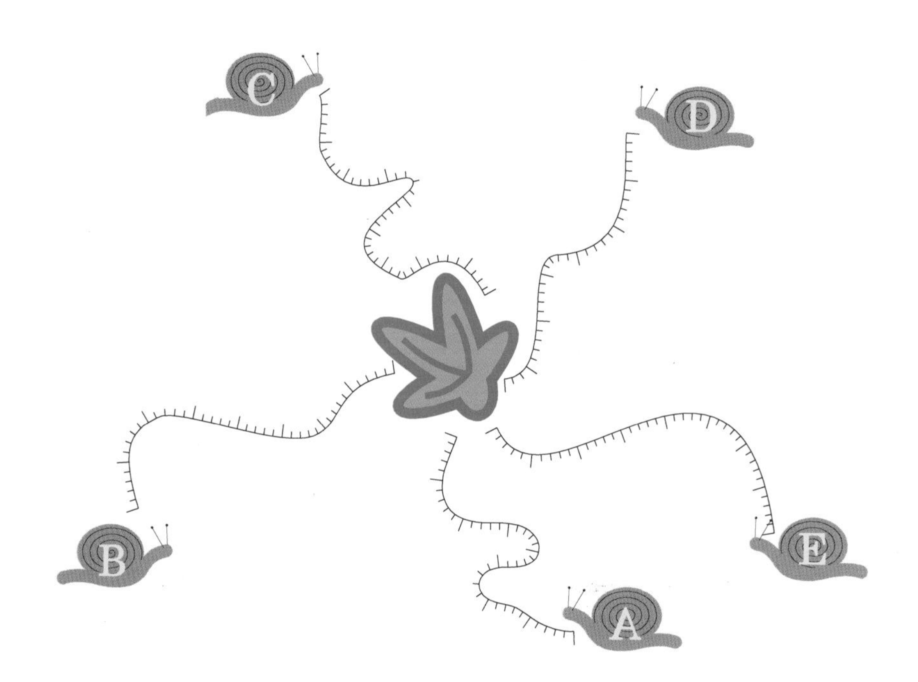

   (A) A **(B)** B **(C)** C **(D)** D **(E)** E

6. The day before yesterday was a Sunday. What day will it be the day after tomorrow?

   (A) Friday **(B)** Monday **(C)** Tuesday **(D)** Wednesday **(E)** Thursday

7. Daniel has 5 toy cars, Matthew has 3 cars more than Daniel, and Rebecca has 2 cars less than Matthew. How many toy cars do the three children have altogether?

   (A) 10 **(B)** 15 **(C)** 16 **(D)** 18 **(E)** 19

8. Shawn's car can travel 60 miles on one gallon of fuel. Danya's car uses up two gallons of fuel every 80 miles. Danya and Shawn live 60 miles apart. How much more fuel does Danya need to drive to Shawn's place than Shawn needs to drive over to Danya's place?

   (A) 0 gal **(B)** 1 gal **(C)** half of a gal **(D)** 2 gal **(E)** 3 gal

4-point problems

9. If the lizard is twice as fast as the snake, and the snake is twice as fast as the tortoise, in which order will the animals reach the finish line?

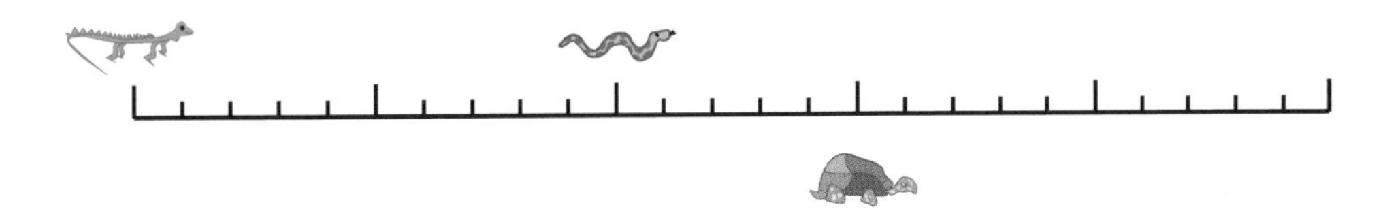

   (A) snake, lizard, tortoise
   (B) lizard, snake, tortoise
   (C) tortoise, snake, lizard
   (D) snake, tortoise, lizard
   (E) lizard, tortoise, snake

10. If 5 numbers have an even sum, then at most how many of them could possibly be odd?

(A) 0 **(B)** 1 **(C)** 3 **(D)** 4 **(E)** 5

11. Antonio has a flat tire in Flatville. There is no mechanic in Flatville but there is a mechanic in each of the other towns on the road sign below. In which of the neighboring towns is the closest mechanic?

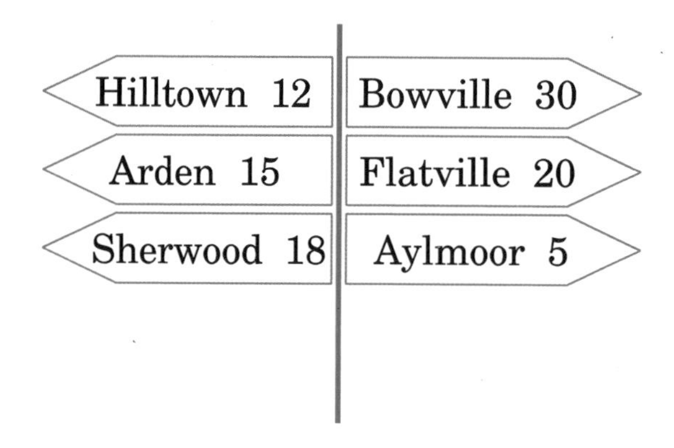

(A) Hilltown **(B)** Bowville **(C)** Arden **(D)** Shwerwood **(E)** Aylmoor

12. The smallest whole number we must add to 50 in order to obtain a result larger than 80 is:

(A) 21 **(B)** 30 **(C)** 31 **(D)** 32 **(E)** 40

13. 4:40 PM was two hours before the start of the class that we arrived 20 minutes late for. What time was it when we arrived at the class?

(A) 2:20 PM **(B)** 3:00 PM **(C)** 6:20 PM **(D)** 7:00 PM **(E)** 7:20 PM

14. Four strings that cannot stretch have been attached at one end, like in the figure. The length of each string is the one marked on the figure. We pull two of the ends A,B,C,D and we measure the distance between the ends. Which distance is longer?

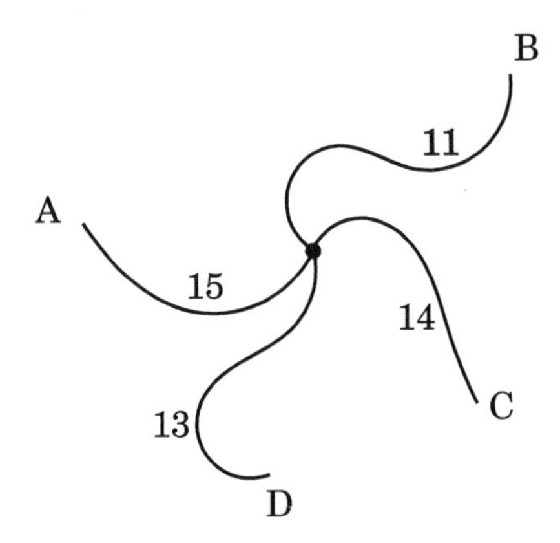

(A) A to B
(B) B to C
(C) C to A
(D) A to D
(E) B to D

15. How many of the figures cannot be cut into 4 identical squares?

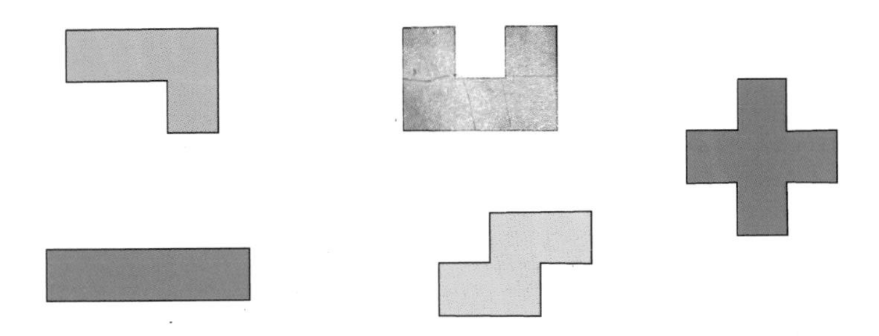

(A) 0 **(B)** 1 **(C)** 2 **(D)** 3 **(E)** 4

16. What is the digit covered by the square?

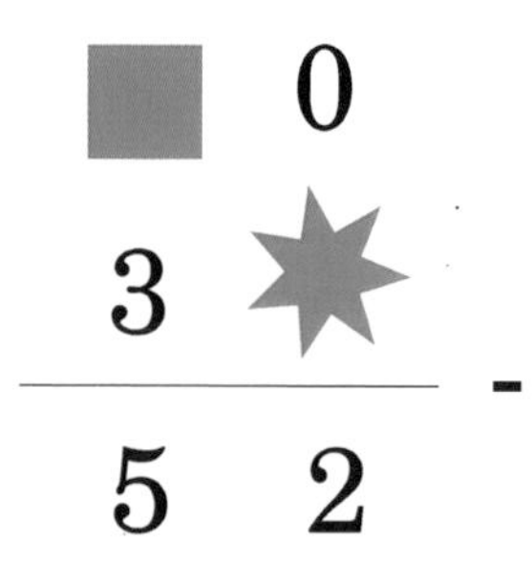

(A) 5 **(B)** 6 **(C)** 7 **(D)** 8 **(E)** 9

5-point problems

17. Jeremy goes skiing every Monday and he goes skating every Sunday. Last month, Jeremy went skiing 4 times and went skating 5 times. The first day of last month could have been a:

(A) Friday **(B)** Monday **(C)** Tuesday **(D)** Wednesday **(E)** Thursday

18. Alice has cut 2 strings in half and 1 string in three parts. How many strings could she have ended up with?

(A) 3 **(B)** 4 **(C)** 5 **(D)** 8 **(E)** 12

19. A 2-digit number has different digits. Another 2-digit number has the same digits as the first one, but in reverse order. What is the largest possible difference between such numbers?

(A) 1 **(B)** 9 **(C)** 64 **(D)** 73 **(E)** 82

20. Four chips have different sizes and colors. The largest chip is not green. The smallest chip is not yellow. The green chip is larger than the yellow chip. The blue chip is larger than the red chip. If we order the chips by size, with the largest on the left and the smallest on the right, which is the correct ordering?

(A) BGYR **(B)** RGYB **(C)** GRYB **(D)** GBRY **(E)** BYRG

21. Jack has two more gold coins than Jill. Jill has 3 more silver coins than Jack. Jill has as many gold coins as Jack has silver coins. Jack has 2 silver coins. How many coins do they have in total?

(A) 7 **(B)** 9 **(C)** 11 **(D)** 13 **(E)** 15

22. If gear W makes one complete turn, how many complete turns does gear M make?

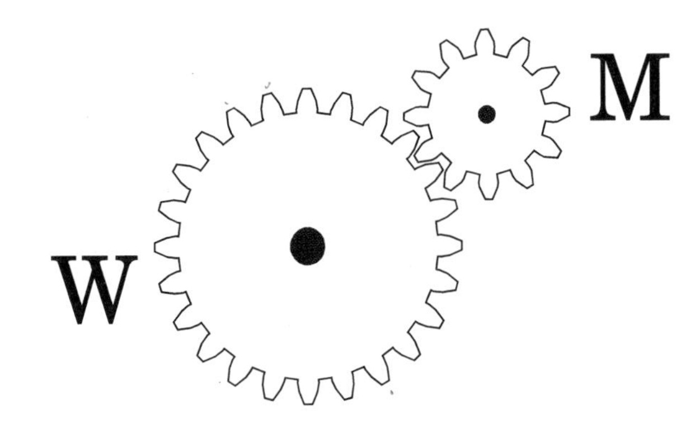

(A) 1 **(B)** 2 **(C)** 3 **(D)** 4 **(E)** 5

23. In the MathRocks band's show, for each song, a different performer plays solo while the others accompany. Jillian, a member of the MathRocks band, has accompanied a 7 songs during the show. How many members are there in the band?

(A) 5 **(B)** 6 **(C)** 7 **(D)** 8 **(E)** 9

24. On planet Rocknroll days are 32 hours long, of which 16 are in daylight and 16 are in darkness. If it is now 9 PM, what time will it be 8 hours from now?

(A) 1 AM **(B)** 2 AM **(C)** 16 PM **(D)** 17 PM **(E)** 18 PM

Answer Key for Test Six

3-point problems		4-point problems		5-point problems	
1.	B	9.	D	17.	A
2.	C	10.	D	18.	D
3.	A	11.	B	19.	E
4.	C	12.	C	20.	A
5.	D	13.	D	21.	D
6.	E	14.	C	22.	B
7.	E	15.	C	23.	D
8.	C	16.	E	24.	A

Hints and Solutions for Test One

1. Make a list:

1st	2nd	3rd	4th	5th	6th	7th	8th	9th	10th	11th	12th
39	38	37	36	35	34	33	32	31	30	29	28

2. The horse will need the same number of steps whether it has the man on its back or not.

3. The fruit can be recognized as: orange, apple, cherry, banana, and watermelon. In alphabetical order they are: apple, banana, cherry, orange, watermelon. The cherry and the watermelon have not changed place.

4. Since one hour has 60 minutes, from 7:35AM it takes 25 minutes to be 8:00AM. After 10 more minutes, it will be 8:10AM.

5. Work backwards: the number smaller than 125 by 11 is 114. The number larger by 7 than 114 is 121.
   Otherwise, since 7 is 4 smaller than 11, just subtract 4 from 125.

6. Notice that the butterflies have the numbers 1, 2, 3, 4, and 5, while the flowers have the numbers 5, 10, 15, 20, and 25. That is, the flowers are counted by fives. The matches are:

1	2	3	4	5
5	10	15	20	25

and 3 corresponds with 15.

7. Make a timeline:

before yesterday	yesterday	today	tomorrow	after tomorrow
Thursday	Friday	Saturday	Sunday	Monday

8. Let us say P represents purple and G represents gold. The visible portion of the pattern is: PGGGGPGGGPGGPGPGGP, where the number of gold stars flanked by two purple stars is decreasing from 4 to 1 after which it looks like it will be increasing back from to 4. The next three objects in the pattern are likely to be GGP.

9. Use *reduction to unity*: a whole bar is used in two weeks, two bars are used in a month. Therefore, 12 bars are used in 6 months.

10. $4 + 8 = 12$

11. Remove block K to find that: $31 - 9 = 22$, which is correct.

12. 20 cents are ten times less than 2 dollars. Therefore, they pay for ten times less balloons.

13. Try to use a piece of string to model the loops.

14. Try to model each operation using paper and scissors.

15. Two-digit numbers start at 10 and end at 99. Since we have to count both 10 and 99 among such numbers, the total number of numbers is $99 - 10 + 1 = 90$.

16. Try to make a model using matches.
17. Since 15 and 14 differ by 1, there is no possibility to subtract a 1 from the sum.
18. Work barckwards:

	Terence	Donald
After Donald gave 6 toys to Terence	13	13
Before Donald gave 6 toys to Terence	7	19
Before Terence gave 8 toys to Donald	15	11

19. A possible coloring:

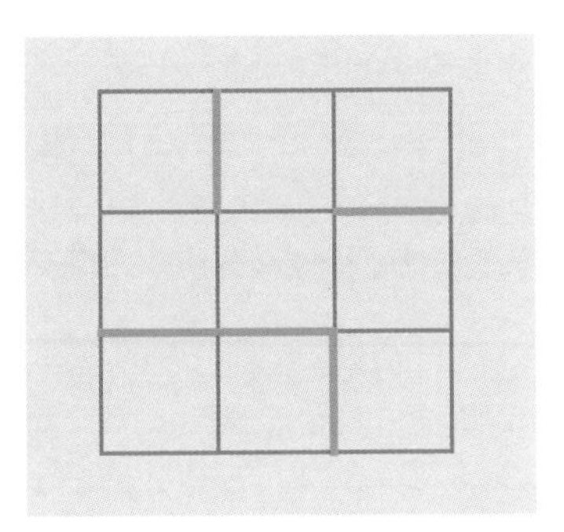

20. If there are no ties, a total of 9 games were played. With two players per game, a total of 18 players participated.

21. Draw shortest paths from the cat to each mouse and count the steps:

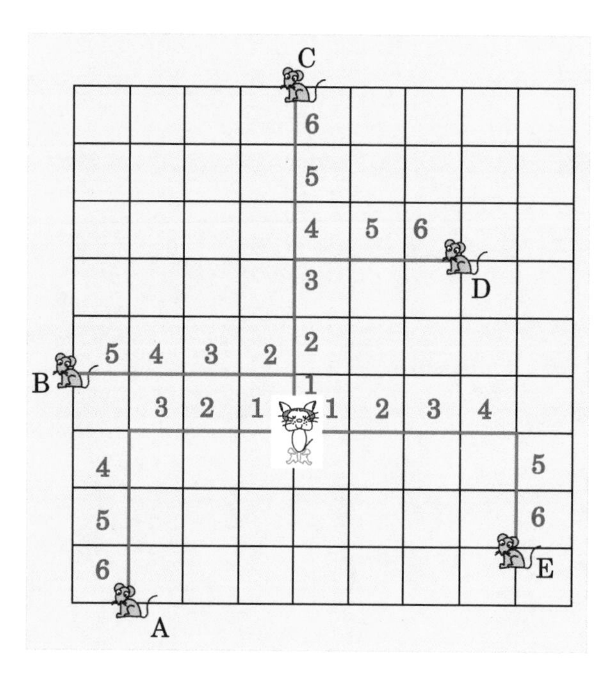

22. While Anda and Stella move, they face each other for half of the time while, for the other half of the time, they are back to back. Therefore, they can see each other for approximately half the time.

23. The middle floor of an 11-story building is the 6th floor. If Erica is on the second floor, the elevator has to go up:

- from second to third
- from third to fourth
- from fourth to fifth
- from fifth to sixth

24. If Lonely and Valentine run at half the speeds everything happens twice slower, including the time lapse between their arrivals to the finish line.

Hints and Solutions for Test Two

1. Cross out the chopsticks in each column and count one chopstick for each of the three plates in a column.
2. Complete the square to find out which is the missing shape:

3. Alena can write a different even digit on a green card as well as a different even digit on a red card and all the cards would be different. Since there are 5 different even digits (0, 2, 4, 6, and 8,) Alena can make 10 different cards.
4. Make sure you do not count the airplane among animals. Also, make sure you count two pairs of wings for the dragonfly and one pair of wings for the bat and the bird. Moreover, notice that you have to count *pairs*, not individual wings.
5. The pattern starts as: R, B, F, R, B. You have to continue it until you find the 11th element: R, B, F, R, B, F, R, B, F, R, B.

6. The sum of two digits can be at most 18, if both digits are equal to 9. However, Alan pressed two keys *at the same time*, which means the digits must be different. If the digits must be different, then they can be at most 9 and 8, with a sum of 17.

7. Make a table of rainy days. If "It has rained yesterday," is true on any Thursday and Sunday, then it rains any Wednesday and Saturday. If "It will rain tomorrow," comes true on any Monday or Thursday, then it rains on any Tuesday and Friday.

Sunday	Monday	Tuesday	Wednesday	Thursday	Friday	Saturday
?	?	Rain	Rain	Rain	Rain	

Only Sundays and Mondays can be sunny. Of these, only Sunday is an answer choice.

8. Be careful not to cross diagonally, as it is not one of the allowed moves. Krista's path is:

9. There are 10 people in total and everyone holds hands with someone. Therefore, there are 10 pairs of hands.

10. To avoid confusion, make a table of the time zones:

Eastville	Midville	Westville
3:10 PM	5:10 PM	6:10 PM
?	?	6:15 AM

Complete the table to find out that it is 3:15 AM in Eastville.

11. There are two possibilities: either I am male or female. If I am male, I would be a brother. Since each sister has two brothers, then there is only one other brother. But then, there are only three sisters and each brothers cannot have four sisters contradiction. Therefore, I am female and there are three more sisters other than myself.

12. Pippi has 8 red socks and 6 blue socks. She can make at most 6 pairs with one blue and one red sock.

13. In each family there are two chicks and one mother - they will need a total of three worms. Each father bird brings three worms, in total 9 worms.

14. Make a useful diagram:

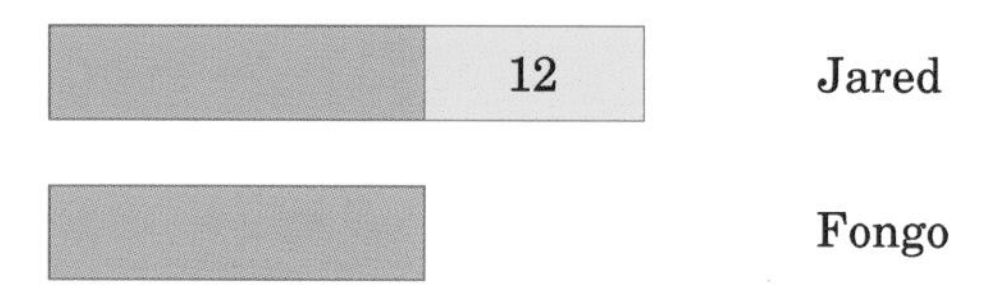

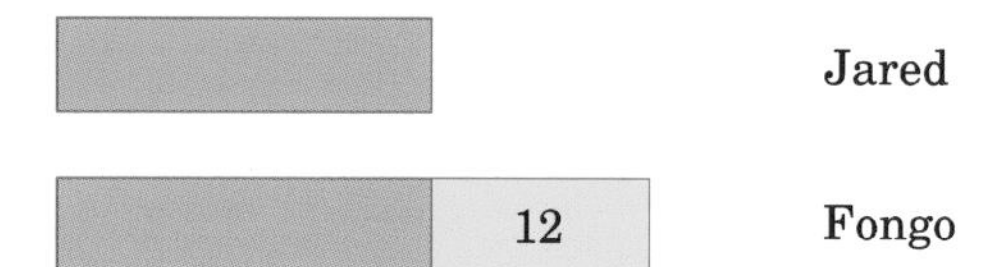

15. Simpleton has 7 golden eggs over the week. At the market, he exchanged 4 eggs for 8 chicks. He had 3 more eggs remaining, for which he got 9 ducklings.

16. Make a timeline:

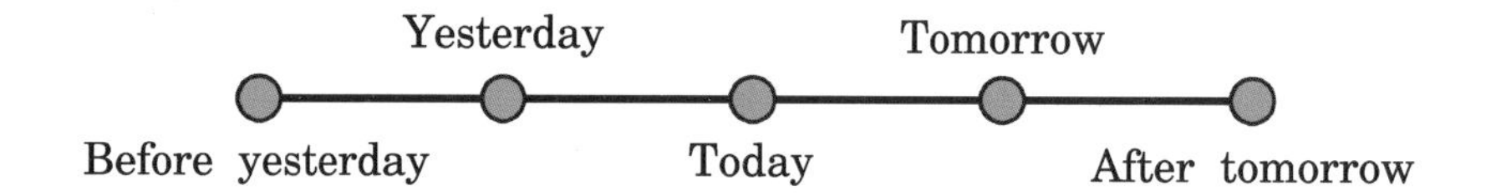

17. Steven added two whole numbers and got 2015.

    - Since 2015 is odd, Steven must have added one even number with one odd number. One of the numbers must be even.
    - Any two numbers can be ordered. Therefore, one of the numbers must be smaller than the other.
    - Since $2015 = 1008 + 1007$, it is not necessarily true that one of the numbers must be smaller than 1000.
    - Since $2015 = 2014 + 1$, it is not necessary for both numbers to be larger than 1000.

18. Make a diagram! In the following diagram the cheese is purple and the standard weight is green:

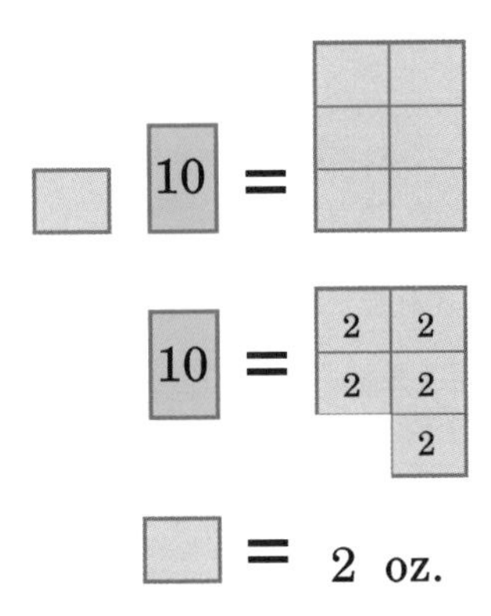

19. Do the same as Jiminy and write some of the alphabet backwards:

 Z Y X W V U T

 until you see the first vowel, which is U. It is the sixth in the sequence of letters.

20. Order the girls by the time it takes each to finish the book. Do this as you read the statement of the problem:

 A T J G S

21. Say the locks are A and B, while the keys are 1, 2, and 3.
 If Akhil starts with key '3', he has to try it in both locks to make sure it is the one that does not fit - 2 total trials. If Akhil starts with key '2' and tries it in lock A, he also has to try it in lock B. He will see that it fits. He will then try one of the remaining keys in lock A. If it fits, then it is key 1 and key 3 is the one he has not tried at all. In this case, 3 total trials.
 Other cases are similar to these.

22. Daniel will be able to form only the words: ABABA and BAAAB.

23. It is not possible to obtain the same result when adding as when subtracting, unless one of the numbers is zero. Since zero is even, choice A is correct.

24. To reverse the order, the girl in the middle stays in the same position. The two girls that are farthest must swap positions, as well as the two girls who flank the girl in the middle:

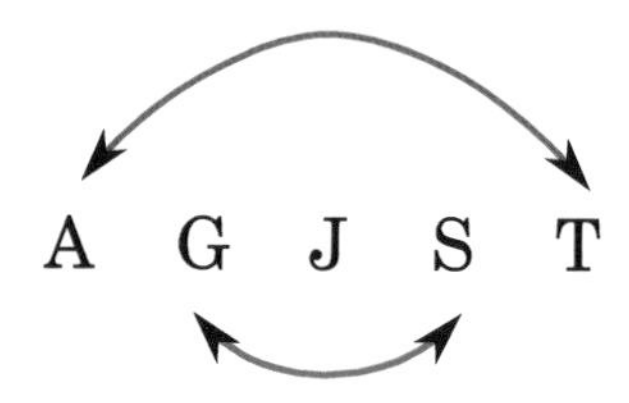

Hints and Solutions for Test Three

1. Underline the number of occurences of the letter A and count them.

2. Count the eggs in each basket. You do not need to count them in any particular order. Start with the basket that looks like a good guess. As soon as you have found the correct basket, stop counting - there is no need to know how many eggs there are in the other baskets.

3. Cross out the trees that are on Fox's path and count the trees that are not crossed out.

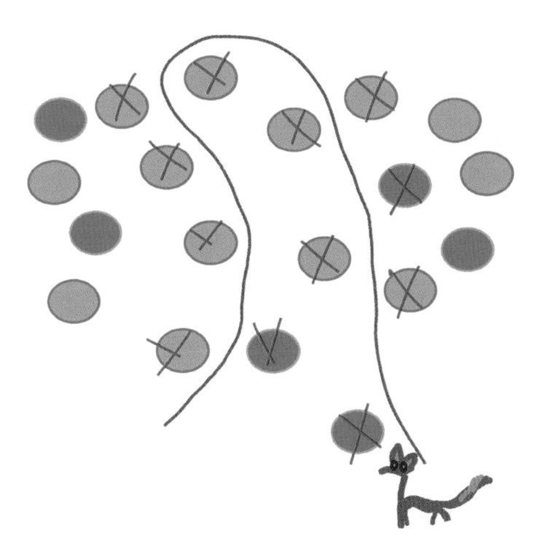

4. Continue the sequence by adding 7 to the last term in order to obtain the next term: 7, 14, 21, 28, 35.

5. Count the number of houses with even numbers - there are 3, therefore he puts 9 letters in their boxes.
   Count the number of houses with odd numbers - there are 4, therefore he puts 8 letters in their boxes.
   The total number of letters is $9 + 8 = 17$.

6. Since 15 minutes represent one quarter of an hour, the car that travels 40 miles in 60 minutes, travels 10 miles in 15 minutes.

7. The cats can have the same two mice as friends.
8. The cards must appear with the yellow card at the top and on the left. Experiment with some cards and a mirror.
9. Harry needs 12 more squares:

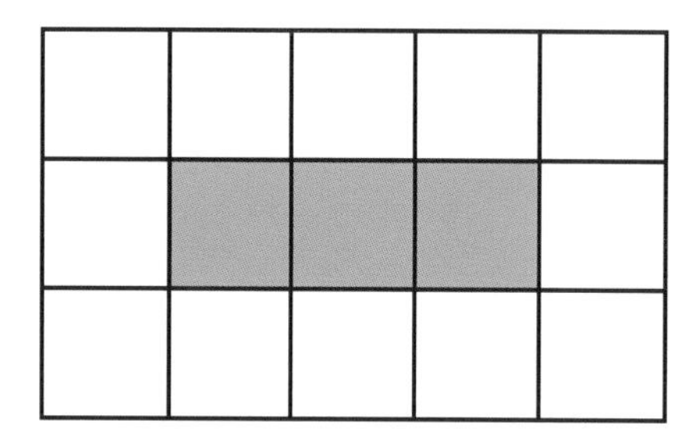

10. If one hour passes, it is 4:00 PM. Subtract 35 minutes from 4:00 PM and obtain 3:25 PM as the time when the class starts.
11. Place the data in a table and perform the same operations as Stan:

Red	Yellow	Green	Operation
5	8	9	Start.
3	8	11	Paint 2 red triangles green.
7	4	11	Paint 4 yellow triangles red.
7	15	0	Paint all the green triangles yellow.

12. Count the highlighted segments for each of the digits. Select the ones that consist of 5 segments only and count them.

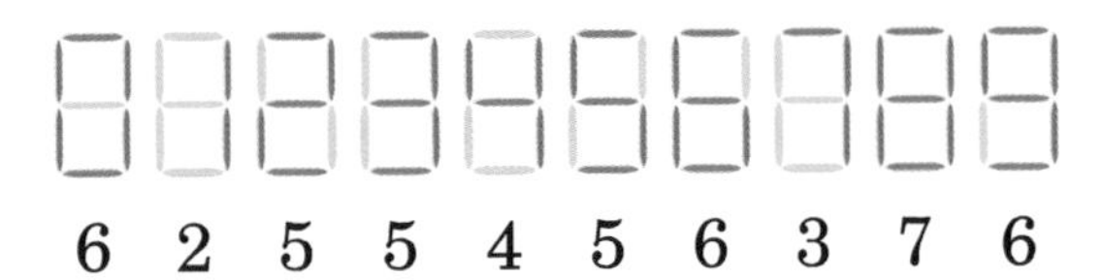

13. Make a weekly timeline and mark Yolanda's return times on it:

Mon	Tue	Wed	Thu	Fri
4:00	4:00	2:30	2:30	4:00

From the timeline, we can see that today can only be Tuesday. The day before yesterday was Sunday.

14. Don't forget that zero is even!

$$
\begin{aligned}
12 &= 12+0 \\
12 &= 10+2 \\
12 &= 8+4 \\
12 &= 6+6
\end{aligned}
$$

15. Subtract 7500 from 9000 to find the difference in altitude.

16. Both operations I did have the outcome that the number decreases by 10. Andrea's operations also have the outcome that the number decreases by 10. Therefore, our results must be identical.

17. Every time George cuts a head, the total number of heads increases by 1 (one head gets cut off and two more grow back.) After 11 cuts, the numer of heads has increased by 11. Add this to the initial number of heads, which is 3, to find the total number of heads.

18. If we subtract 2 from 4 we get 2 again. However, in this case the triangle is 2 and cannot also be 8 or 6. Therefore, the triangle cannot be 2.
Another possibility for the triangle is to be 7, if we borrow from the next larger place value. Since $14 - 7 = 7$, we replace the triangle by a 7. This matches the remaining operation, since we borrowed from the 8 already.

19. Work backwards:

Andy	Brad	Dan	Operation
4	4	4	End.
5	4	4	Dan ate one of Andy's biscuits.
5	4	6	Brad ate 2 of Dan's biscuits.
5	7	6	Andy ate 3 of Brad's biscuits.

Brad had 7 biscuits to start with.

20. Make a diagram with the coconuts:

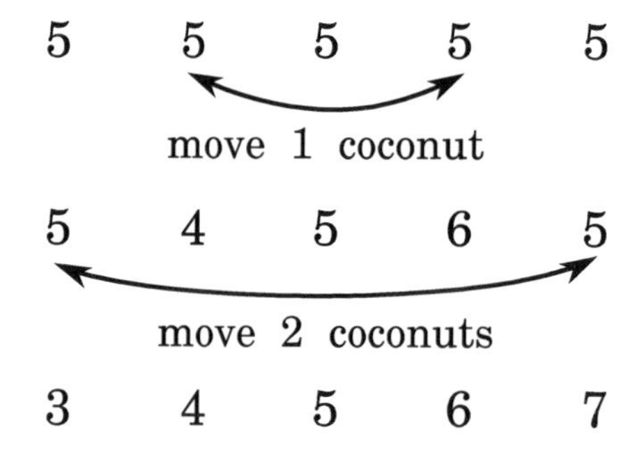

21. Any digit sum of a 2-digit number is smaller than or equal to 18. For the difference of two digit sums to be 17, the sums must be one of:

$$
\begin{aligned}
17 &= 18 - 1 \\
17 &= 17 - 0
\end{aligned}
$$

 Since the first digit sum is even and the second is odd, we must be in the case that $17 = 18 - 1$. Only the 2-digit number 99 has a digit sum of 18 and, therefore, the 4 digit number in question starts with a 9.

22. If 3 glooks weigh as much as 2 maboons, they weigh as much as 9 pogs. Therefore, 1 glook weighs as much as 3 pogs.

23. There are 6 total ways in which Anna, Chris, and Donald can stand in a line:

 ACD, ADC, DAC, DCA, CAD, CDA

 Of these, one must be the ordering by increasing height and one must be the ordering by decreasing height. Therefore, there are 4 ways they can stand in a line without being ordered by height.

24. Experiment with an object that has the same general shape as a cube, such as a thick book. Place a toy on any of the faces and make it go only forward. You will notice that it can go either 3 steps (if the cube sits on a table) or 4 steps, since it cannot change direction. At most, a toy can get 4 stamps in its passport.

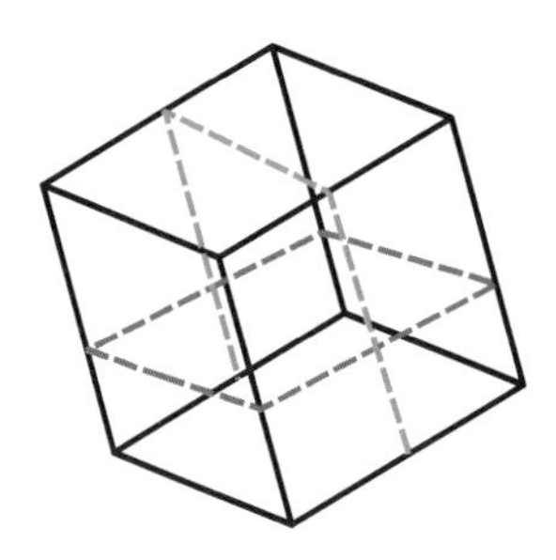

Hints and Solutions for Test Four

1. Since $6+6=12$ and $12-11=1$, the result must be 2.

2. Count the dots on each balloon and compare:

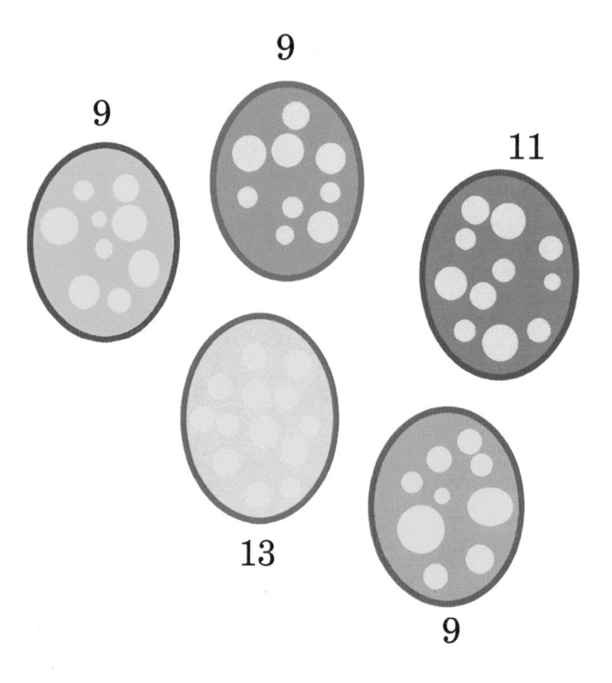

3. Each person has two boots. There are 4 people, therefore there are 8 boots.

4. Each figure in an even position is a rotated version of the figure in the odd position immediately on its left. The rotation is clockwise by a quarter circle.

5. In any month there are at most three days of the week that can occur five times. These days are next to each other. Therefore, if there are 5 Thursdays, there can also be 5 Tuesdays, 5 Wednesdays, 5 Fridays, and/or 5 Saturdays.

6. We can use three strokes that cut the cake from side to side and do not intersect all at the same point:

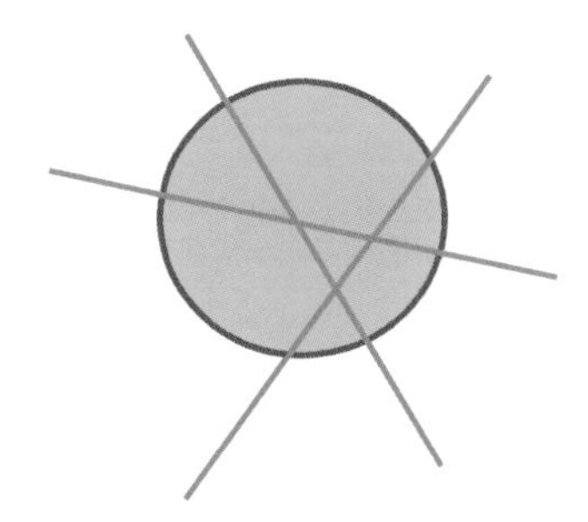

7. Andra can stamp all the squares first and all the triangles after. In this way, she changes the stamp she is holding only one time.

8. Jan can use sticks of the same length or sticks of different lengths to make the side of a square. The possible sizes are: $4+4$, $6+6$, $3+3$, $4+3$, $4+6$, and $3+6$:

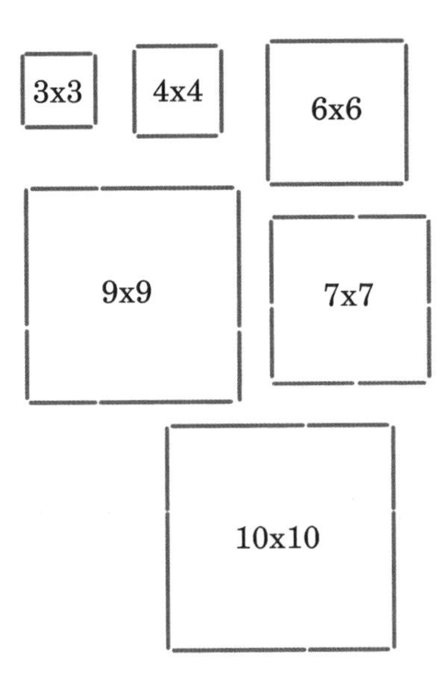

9. Model the problem using a sheet of paper instead of a blanket.

10. Each part of the train travels the same distance!

11. Based on their statements, the numbers are:

    (a) - One less than four pairs is $8 - 1 = 7$.

    (b) - The largest digit is 9.

    (c) - The an even digit larger than 6 is 8.

    (d) - Three more than two pairs is $4 + 3 = 7$.

    (e) - The second largest odd digit is 7.

    There are 3 different digits.

12. Make a diagram:

    10th

    9 cars

    12 cars

    13th

    The total number of cars in the row is: $12 + 1 + 9 = 22$, since our car should only be counted once.

13. The farmer has to weigh 2 lbs of cherries and set them aside. Then, he has to weigh another 2 lbs. Lastly, he has to make one more measurement, placing the cherries from the measured 2 lbs on both pans until he obtains two piles that weigh the same. He will combine the cherries from the first measurement with the cherries on one pan from the third measurement.

14. Some daughters could also be mothers. One possible diagram for the least number of people in the group is:

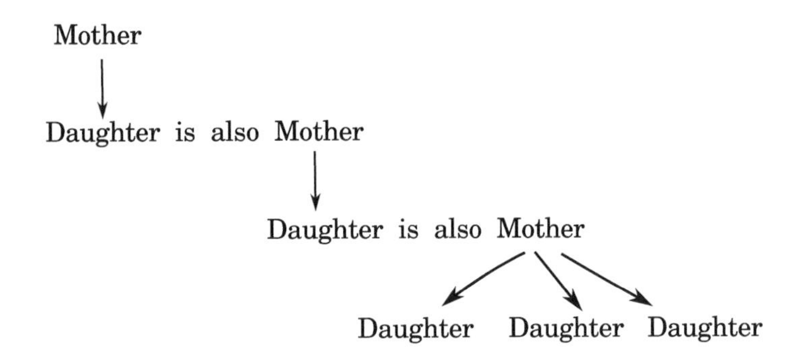

Other diagrams are possible, but they all have at least 6 people.

15. After the small circle does half a turn clockwise and the large circle does half a turn counter-clockwise, the arrow marked 4 points to the same figure as before the rotations.

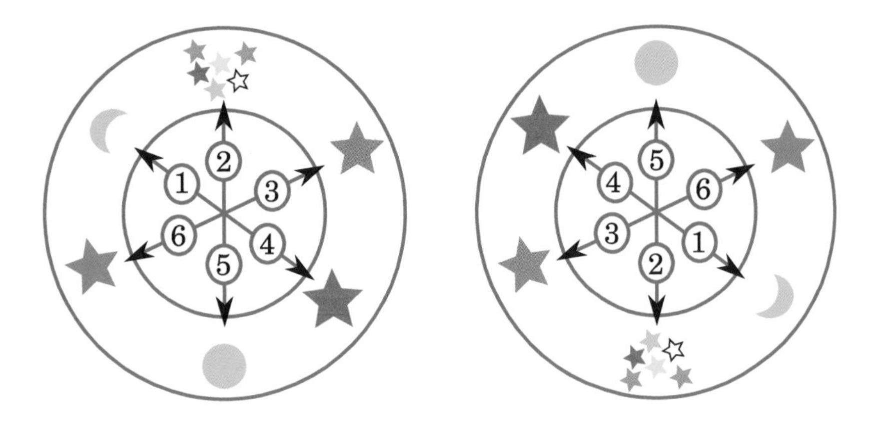

16. The color of the card changes according to the sequence:

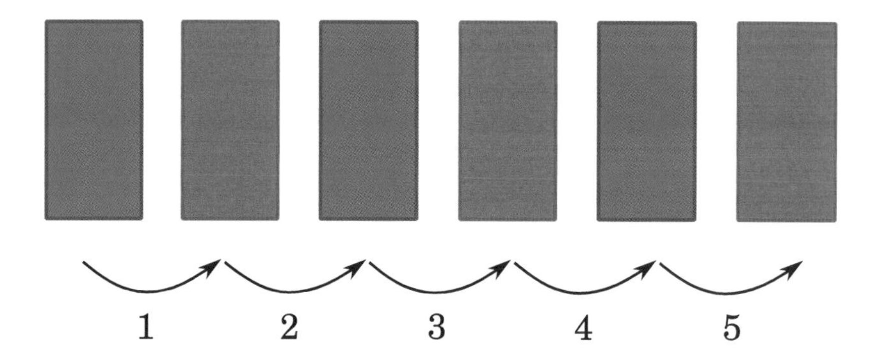

The arrows label the number of turns.

17. Two of the statements are true:

    1. True, because all the green balloons are small.
    2. True, because all the large balloons are orange.
    3. False, because there is a small balloon that is orange.
    4. False, because there is an orange balloon that is small.

18. Make a diagram. From the diagram, we see that half of Shawn's toys must be equal to 3 toys. Therefore, Shawn has 6 toys. This means that Mark has only 1 toy and Daniel has 3 toys.

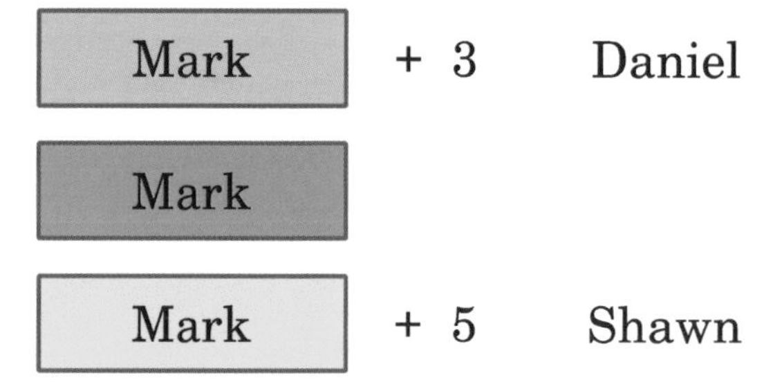

19. Experiment with the digits to figure out how to make the difference smallest. If one number is 789 and the other is 787, the difference is 2.

20. To make a tower out of 5 cubes, Dan must glue 3 pairs of faces together. The three cubes in the middle must have at least two blue faces:

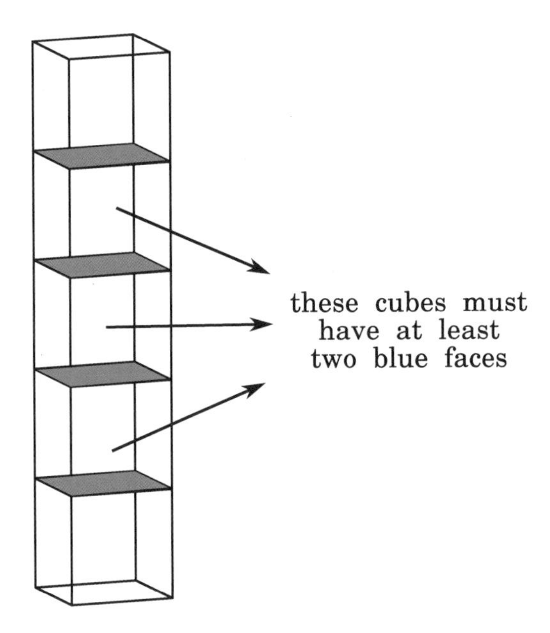

21. Make a diagram that represents all the information in the statement. Notice that, if Papa bear gives one small rectangle to Little Bear, then they each have 2 little rectangles. This means that each rectangle represents 3 apples.

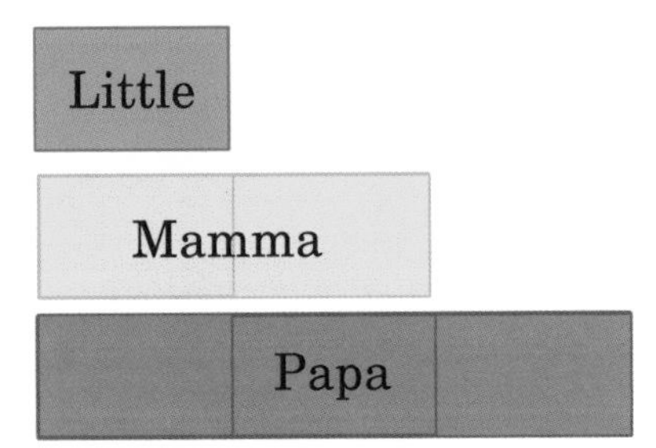

22. Laura can use the cards to make any of the numbers: 6, 9, 66, 99, 69, and 96.

23. There will be as many layers of paper as small rectangles there are. Each rectangle forms a layer.

24. Notice that each exchange involves two animals at the same time. If the Bear exchanged 3 cards, he must have exchanged with each of: Wolf, Fox, and Rabbit. Already these three animals have one card exchange made. Since Rabbit exchanged only one card, this is the only exchange for Rabbit. If Wolf exchanged 2 cards, one of them has surely been exchanged with Bear and the other cannot have been exchanged with Rabbit. Therefore, it must have been exchanged with Fox. Fox cannot exchange a card with Rabbit either and therefore, Fox must have exchanged only 2 cards.

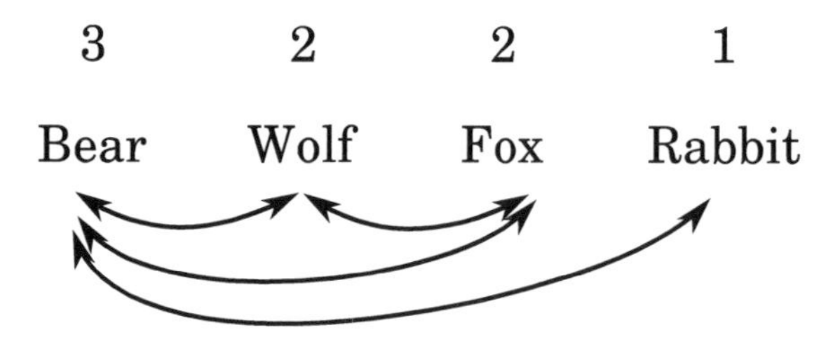

Hints and Solutions for Test Five

1. Count the petals and the leaves on each flower:

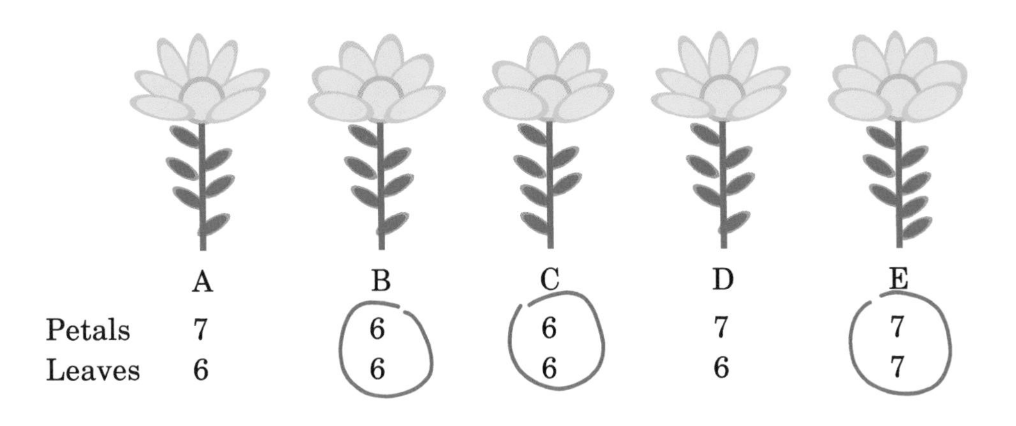

2. Count the vowels in each word and select the words with 3 vowels:

TREE 2 LEAF 2 PEAR 2 DOUBLE 3 FAREWELL 3

FLAMINGO 3 QUINCE 3 HORSE 2 ROSIN 2 MICE 2

3. Since one hour has 60 minutes, there are 12 groups of 5 minutes in one hour. It will take 12 days for the sun to rise 1 hour earlier.

4. Be sure to notice that there are 2 socks to a pair but also 'a pair of pants' is, in fact, only one garment. John has to change: $2 + 1 + 1 + 1 + 1 = 6$ pieces of clothing.

5. Complete the drawing *with a different color pencil.* In this way, you will not miss any of the triangles that have to be counted:

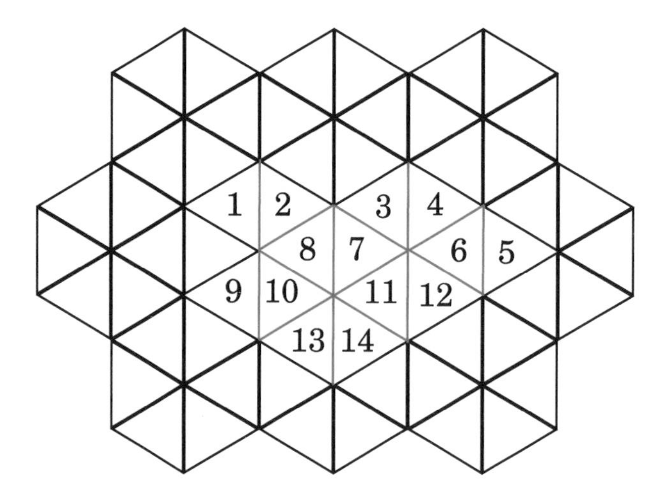

6. Make a diagram to keep track of the number of animals:

3	W		W		W	
3	B		B		B	
6	C	C	C	C	C	C
6	D	D	D	D	D	D
+ 18						

There are 18 animals.

7. To chime 7 times, the clock will chime ding-ding-dong-ding-ding-dong-ding. The number of seconds it takes for the chimes is: $1+1+2+1+1+2+1=9$ seconds.

8. Notice that, Alice, Ben, Chloe, and Daniel, are only four people. There must be another person among the five, and this is a person who left with their own umbrella.

9. The number of points accumulated by Andra is:

$$3\times 4+2\times 3=18$$

   and the number of points accumulated by Tom is:

$$4\times 2+1\times 6=14$$

   The difference in scores is 4 points.

10. Figure out which square obstructs another.

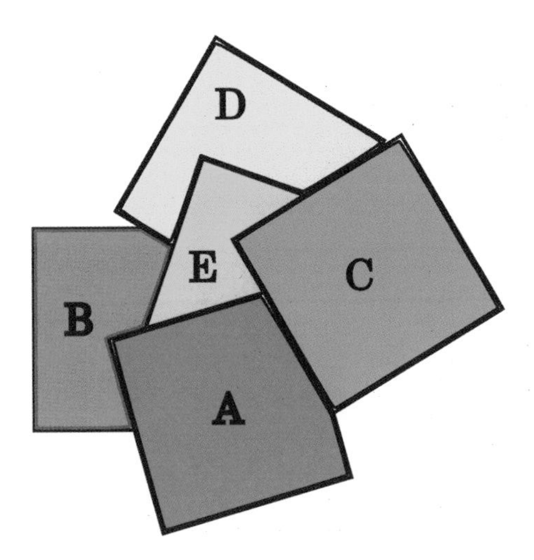

11. Make a diagram with Kro's and Kra's movements. Notice that if Kro takes 4 leaps in Kra's direction and Kra takes 4 leaps in Kro's direction, the two frogs merely exchange their positions.

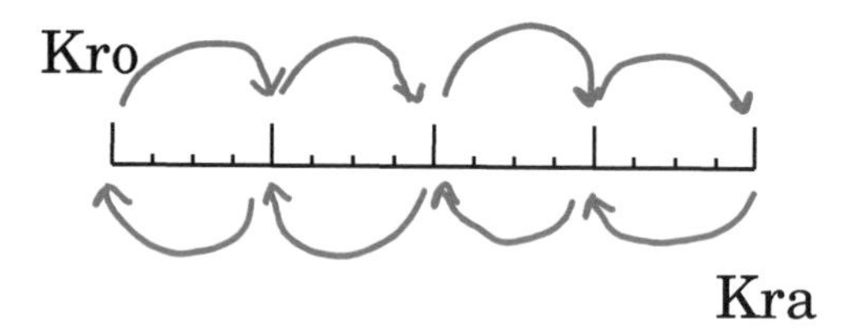

12. Figure out how the pattern works by inserting the missing days. Notice that Santa goes to the workshop every fifth day:

Thursday, *Friday, Saturday, Sunday, Monday,* **Tuesday**, *Wednesday, Thursday, Friday, Saturday,* Sunday, *Monday, Tuesday, Wednesday, Thursday,* **Friday**, *Saturday, Sunday, Monday Tuesday,* **Wednesday**, *Thursday, Friday, Saturday, Sunday,* **Monday**

13. Make a diagram to figure out how far apart the two lockers are:

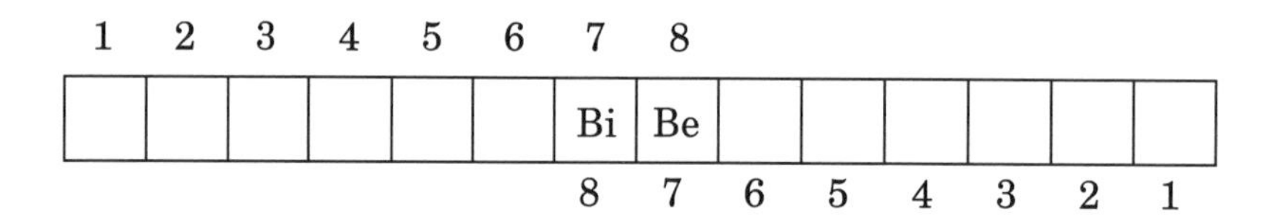

The two lockers are side by side. There are zero lockers between them.

14. For three pounds of grapes all the quantities are three times larger: 2×3 pineapples and 3×3 apples. This corresponds to the formula (A).

15. The rectangle that Hedwig cut is 9×1. The smallest square it can come from is a 9×9 square and Hedwig must have cut it into 9 rectangles.

16. The month of March has 31 days in total. Subtract $12 + 16 = 28$ days to find the number of days with temperatures between 48° F and 50°.

17. The difference $180 - 160 = 20$ lbs represents the weight of the second suitcase. Since the first suitcase has the same weight, Martin must weigh $160 - 20 = 140$ lbs.

18. Make a table of the distances from the top of the tower to the bottom of the cellar for each of the sandcastles:

$$\begin{array}{rcl} \text{Sandeep} & 8+5 & = 13 \\ \text{Alvaro} & 12+1 & = 13 \\ \text{Chloe} & 11+2 & = 13 \\ \text{Martin} & 9+5 & = 14 \\ \text{Leo} & 10+2 & = 12 \end{array}$$

19. There is actually no need to write the numbers in increasing order. It is sufficient to notice that the 2-digit, 3-digit, and 4-digit numbers are not close to 50500, therefore just count them as lower: there are 5 such numbers. Then, start with the 5-digit numbers and order these:

$$50055, 50505$$

Already one can see that 50505 is larger than 50500. Therefore, 50500 must be after 50055.

20. Work backwards:

Operation	Dana	Jack	Telly
End	6	6	6
Jack gave 3 pretzels to Telly	6	9	3
Dana gave 5 pretzels to Jack	11	4	3

21. If the whole performance lasted 8 minutes, and 1 minute was taken by the group performance, 7 minutes were left for individual performances. Therefore, there were 7 pairs who danced. If there are 7 pairs, then there are 14 people

22. To obtain a 4 in the hundreds' place, there must be a carryover of 1 from the tens' place. To obtain the same digit by adding a different triangle to the square, is not possible unless there is a carryover of 1 from the ones' place. Since the yellow circle is the result of 'red square plus green triangle' as well as of 'red square plus 1 plus blue triangle', the two digits covered by triangles must have a difference of 1. Indeed, all the solutions below are correct:

$$
\begin{aligned}
311 + 89 &= 400 \\
322 + 89 &= 411 \\
322 + 78 &= 400 \\
333 + 89 &= 422
\end{aligned}
$$

etc.

23. Draw the paths that seem shortest and count the number of rooms on each path. The shortest paths cross 12 rooms.

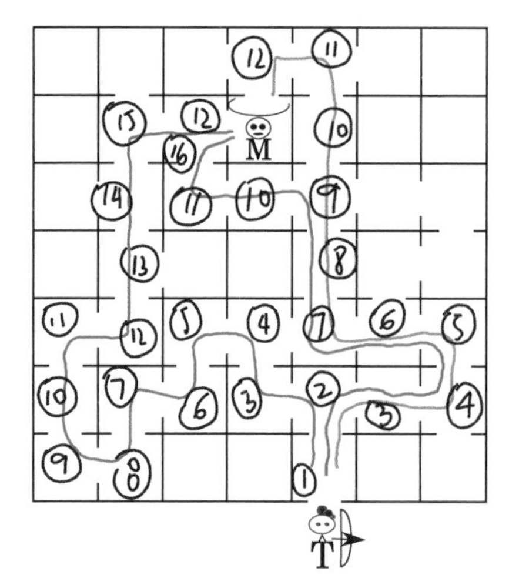

24. Andrew can fit at most 5 white squares. A possible (not unique!) solution is given in the figure:

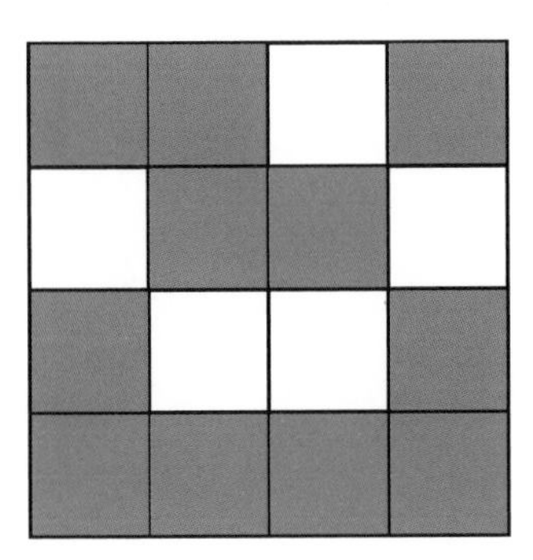

HINTS AND SOLUTIONS FOR TEST SIX

1. Select the numbers starting with '899' since the ones starting with '898' will be smaller than them: 899899 and 899889. Of these, 899899 is larger than 899889.

2. Just add the height of the balloon (altitude) to the depth of the lake: $200 + 50 = 250$ yards.

3. Mark the hours on the diagram:

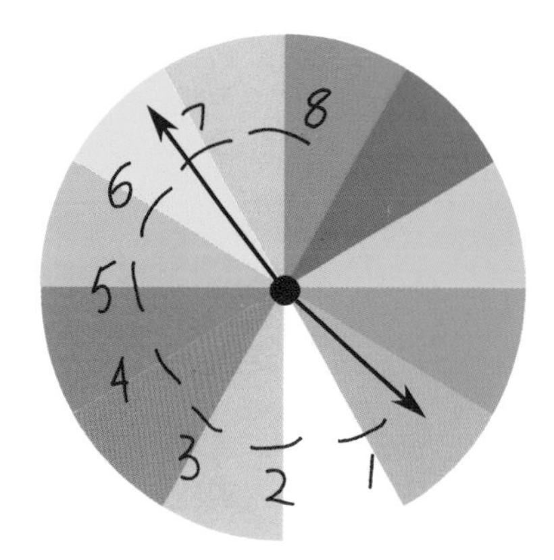

4. Multiply the cost of the ticket by the number of people who pay that price. For the adults: $2 \times 10 = 20$. For the children: $3 \times 5 = 15$. In total: $15 + 20 = 35$.

5. The paths of the snails are marked with ruler markings. Count the number of units on each path and compare:

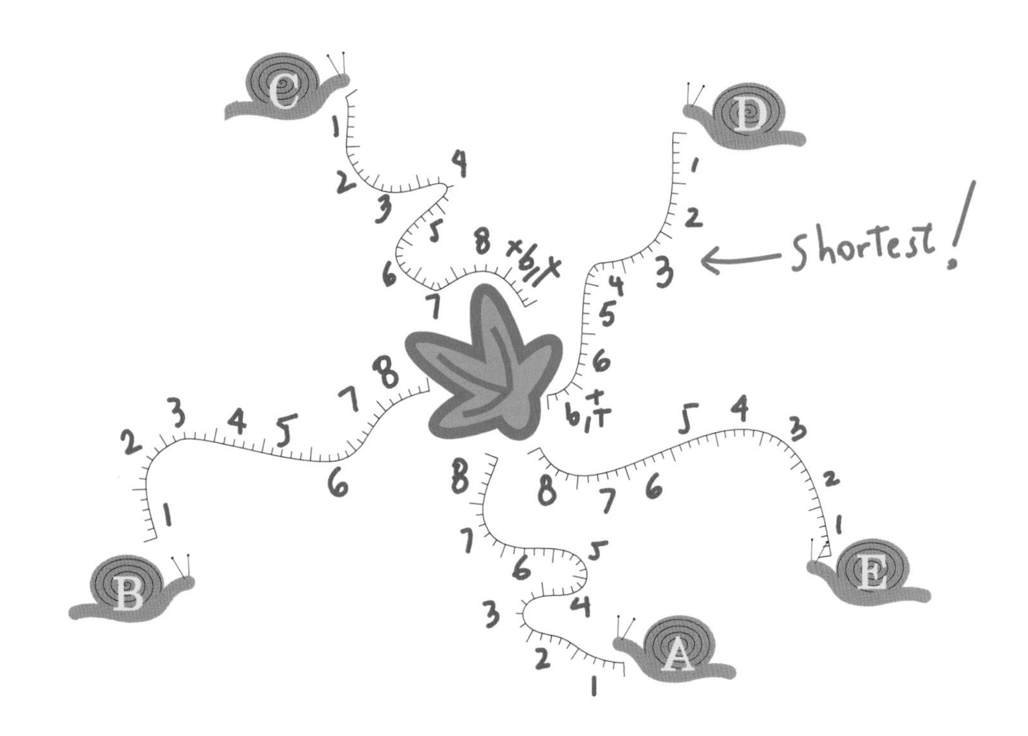

(A) A **(B)** B **(C)** C **(D)** D **(E)** E

6. Make a timeline and place the days on it:

Sunday Monday Tuesday Wednesday Thursday

7. Matthew has $5 + 3 = 8$ cars and Rebecca has $8 - 2 = 6$. In total, the three children have: $5 + 8 + 6 = 19$ toy cars.
   (A) 6 **(B)** 10 **(C)** 16 **(D)** 18 **(E)** 19

8. Shawn's car will use exactly one gallon to travel to Danya's house. Danya's car uses one gallon for 40 miles and another half a gallon for 20 miles. In total, Danya's car will use one gallon and a half to travel to Shawn's house. Danya's car will use half a gallon more than Shawn's car.

9. As the tortoise travels two units as marked on the ruler, the snake travels 4 units. Since the snake only has 3 units to the finish line, it will arrive there before the tortoise.
   As the snake travels 3 units, the lizard can travel 6 units. However, the lizard has only 5 units to go and it will arrive at the finish line before the snake.

10. Two numbers have an even sum if they are both even or both odd. Four odd numbers can be paired together to have even sums to which we can add one even number for an even total.

11. The distances are as follows:

    From Flatville to Bowville there are 10 miles.

    From Flatville to Aylmoor there are 15 miles.

    From Flatville to Hilltown there are 32 miles.

    From Flatville to Arden there are 35 miles.

    From Flatville to Sherwood there are 23 miles.

    The shortest distance is between Flatville and Bowville.

12. Make sure the result is *larger* than 80. Since $50 + 31 = 81$, 31 is the smallest whole number we must add to 50 in order to obtain a result larger than 80.

13. The class must have started at 2:40 PM. Since we arrived 20 minutes later, we must have arrived at 3:00 PM.

14. We must add the lengths of the strings for all of the distances required for comparison:

    A to B is $15 + 11 = 26$.

    B to C is $11 + 14 = 25$.

    C to A is $14 + 15 = 29$.

    A to D is $15 + 13 = 28$.

    B to D is $11 + 13 = 24$

    Compare and select the shortest distance.

15. Mark the squares and count them for each figure:

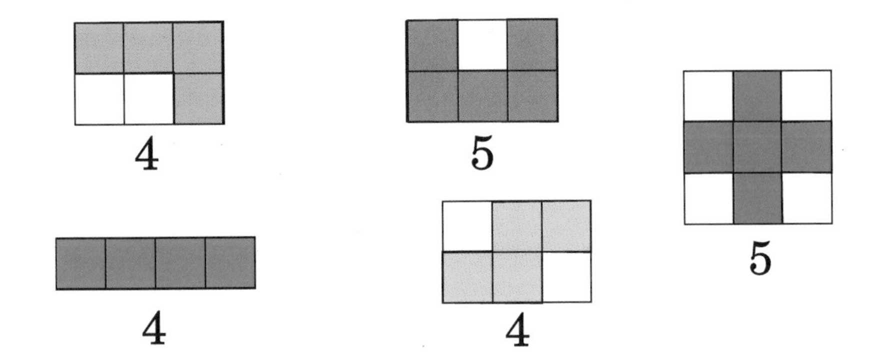

16. The star must be subtracted from 10, since borrowing is absolutely necessary. Therefore, the star must equal 8. To obtain 5 after subtracting 3 from the square, the square must be equal to 9, since we borrowed from it.

17. If Jeremy had an activity 5 times on Mondays and 4 times on Sundays, this means that the last day of the month must have been a Sunday. The Monday following it was already the next month and did not contribute to the count. Since months can have up to 31 days, the first day of the month could have been a: Sunday, Saturday, or Friday. Of these, only Friday is an answer choice.

18. There are several possibilities for picking the specific strings that Alice cut. Make a diagram to represent them:
Suppose Alice started with 3 different strings:

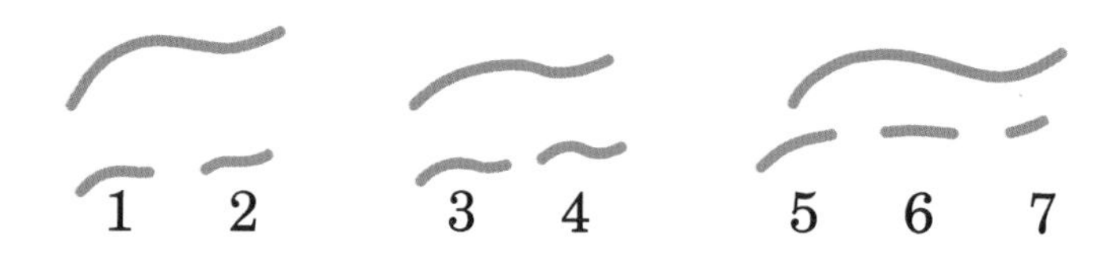

Suppose Alice started with 2 different strings:

Suppose Alice started with 1 string:

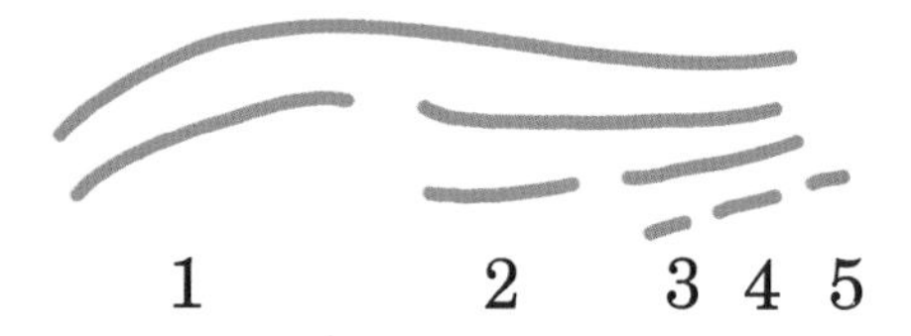

Therefore, Alice may end up with 5, 6, or 7 strings. Of these results, only 5 is an answer choice.

19. For the difference to be largest, the tens digit of the difference must be the largest possible - for this to happen, one of the digits must be 9 and the other digit must be 1:

$$91 - 19 = 82$$

20. Make a diagram with four positions and place the chips according to the description.
    If the the largest chip is not green and the smallest chip is not yellow, the green and the yellow chips must be in the middle positions. Since the green chip is larger than the yellow chip, it must be on the left of the yellow. If the blue chip is larger than the red chip, then the blue chip is in the leftmost position and the red one is in the rightmost.

21. If Jack has 2 silver coins and Jill has as many gold coins as Jack, then Jill must have 2 gold coins. If Jack has 2 more gold coins than Jill, then Jack must have 4 gold coins. If Jill must have 3 more silver coins than Jack, then she has 5 silver coins.
    Jack has 6 coins in total and Jill has 7 coins in total. Together, they have 13 coins.

22. Count the teeth of each gear. Gear W has 24 teeth and gear M has 12 teeth. As gear W turns once, gear M must turn twice.

23. Jillian has accompanied 7 songs and has played a solo. Therefroe, there are 8 members in the band.

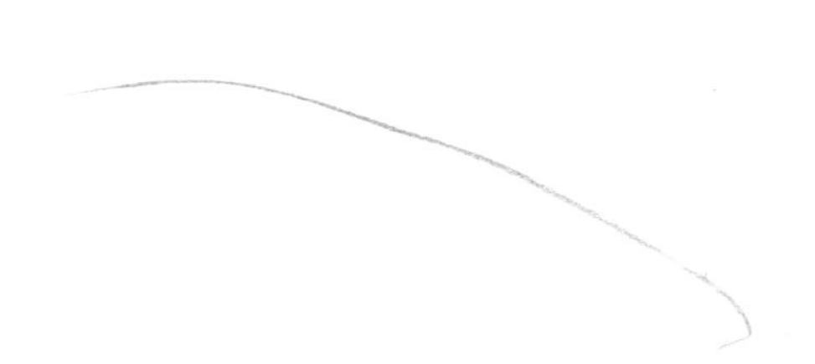

24. Make a clock face divided in 16:

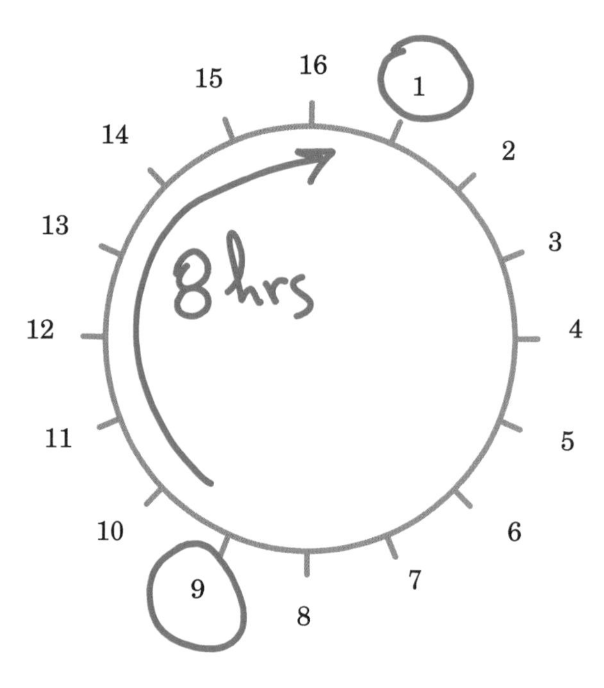

Alternately, add $8 + 9 = 17$ and subtract 16 to find that it is 1AM.